Volkswagen Inspection/Maintenance (I/M) Emission Test Handbook 1980 ▶ 1997

Overview and Troubleshooting
for Volkswagen Cars, Vans,
and Pickups

RB
Robert Bentley
Cambridge, Massachusetts

Volkswagen Service Manuals from Robert Bentley

Volkswagen Inspection/Maintenance (I/M) Emission Test Handbook: 1980-1997 Overview and Troubleshooting for Volkswagen Cars, Vans, and Pickups. Robert Bentley and Volkswagen United States. ISBN 0-8376-0394-3 *Volkswagen Part No. LPV 800 901*

Passat Official Factory Repair Manual: 1995-1997
Gasoline, Turbo Diesel, and TDI, including wagon
Volkswagen United States. ISBN 0-8376-0380-3
Volkswagen Part No. LPV 800 207

Jetta, Golf, GTI, Cabrio Service Manual: 1993-1997 including Golf$_{III}$ and Jetta$_{III}$
Robert Bentley. ISBN 0-8376-0365-X
Volkswagen Part No. LPV 800 114

GTI, Golf, Jetta Service Manual: 1985-1992 Gasoline, Diesel, and Turbo Diesel, including 16V.
Robert Bentley. ISBN 0-8376-0342-0
Volkswagen Part No. LPV 800 112

Corrado Official Factory Repair Manual: 1990-1994
Volkswagen United States. ISBN 0-8376-0387-0
Volkswagen Part No. LPV 800 300

Passat Service Manual: 1990-1993 including Wagon. Robert Bentley. ISBN 0-8376-0378-1
Volkswagen Part No. LPV 800 205

Cabriolet, Scirocco Service Manual: 1985-1993, including Scirocco 16V. Robert Bentley.
ISBN 0-8376-0362-5
Volkswagen Part No. LPV 800 113

Volkswagen Fox Service Manual: 1987-1993,
including GL, GL Sport and Wagon.
Robert Bentley ISBN 0-8376-0363-3
Volkswagen Part No. LPV 800 504

Quantum Official Factory Repair Manual: 1982-1988
Gasoline and Turbo Diesel, including Wagon and Syncro.
Volkswagen United States. ISBN 0-8376-0341-2 *Volkswagen Part No. LPV 800 202*

Vanagon Official Factory Repair Manual: 1980-1991 Including Diesel, Syncro and Camper.
Volkswagen United States. ISBN 0-8376-0336-6
Volkswagen Part No. LPV 800 148

Rabbit, Scirocco, Jetta Service Manual: 1980-1984 Gasoline Models, including Pickup Truck, Convertible, and GTI. Robert Bentley. ISBN 0-8376-0183-5 *Volkswagen Part No. LPV 800 104*

Rabbit, Jetta Service Manual: 1977-1984 Diesel Models, including Pickup Truck and Turbo Diesel.
Robert Bentley. ISBN 0-8376-0184-3
Volkswagen Part No. LPV 800 122

Rabbit, Scirocco Service Manual: 1975-1979 Gasoline Models. Robert Bentley. ISBN 0-8376-0107-X
Volkswagen Part No. LPV 997 174

Dasher Service Manual: 1974-1981 including Diesel. Robert Bentley. ISBN 0-8376-0083-9
Volkswagen Part No. LPV 997 335

Super Beetle, Beetle and Karmann Ghia Official Service Manual Type 1: 1970-1979
Volkswagen United States. ISBN 0-8376-0096-0
Volkswagen Part No. LPV 997 109

Beetle and Karmann Ghia Official Service Manual Type 1: 1966-1969
Volkswagen United States. ISBN 0-8376-0416-8
Volkswagen Part No. LPV 997 169

Station Wagon/Bus Official Service Manual Type 2: 1968-1979 Volkswagen United States.
ISBN 0-8376-0094-4
Volkswagen Part No. LPV 997 288

Fastback and Squareback Official Service Manual Type 3: 1968-1973 Volkswagen United States.
ISBN 0-8376-0057-X
Volkswagen Part No. LPV 997 383

Volkswagen 1200 Workshop Manual: 1961-1965, Type 11, 14 and 15 Beetle, Beetle Convertible, Karmann Ghia Coupe and Karmann Ghia Convertible
Robert Bentley. ISBN 0-8376-0390-0
Volkswagen Part No. LPV 800 121

Volkswagen Transporter Workshop Manual: 1963-1967, Type 2 including Kombi, Micro Bus, Micro Bus De Luxe, Pick-up, Delivery Van and Ambulance.
Robert Bentley. ISBN 0-8376-0391-9
Volkswagen Part No. LPV 800 135

Audi Service Manuals from Robert Bentley

Audi 100, A6 Official Factory Repair Manual: 1992-1997, including S4, S6, quattro and Wagon models. Audi of America. ISBN 0-8376-0374-9
Audi Part No. LPV 800 702

Audi 80, 90, Coupe Quattro Official Factory Repair Manual: 1988-1992 including 80 Quattro, 90 Quattro and 20-valve models. Includes Electrical Troubleshooting Manual.
Audi of America. ISBN 0-8376-0367-6
Audi Part No. LPV 800 604

Audi 100, 200 Official Factory Repair Manual: 1989-1991, including 100 Quattro, 200 Quattro, Wagon, Turbo and 20V Turbo. Audi of America. ISBN 0-8376-0372-2
Audi Part No. LPV 800 701

Audi 5000S, 5000CS Official Factory Repair Manual: 1984-1988 Gasoline, Turbo, and Turbo Diesel, including Wagon and Quattro. Audi of America. ISBN 0-8376-0370-6
Audi Part No. LPV 800 445

Audi 5000, 5000S Official Factory Repair Manual: 1977-1983 Gasoline and Turbo Gasoline, Diesel and Turbo Diesel. Audi of America. ISBN 0-8376-0352-8
Audi Part No. LPV 800 443

Audi 4000S, 4000CS, and Coupe GT Official Factory Repair Manual: 1984-1987 including Quattro and Quattro Turbo. Audi of America. ISBN 0-8376-0373-0 *Audi Part No. LPV 800 424*

Audi 4000, Coupe Official Factory Repair Manual: 1980-1983 Gasoline, Diesel, and Turbo Diesel.
Audi of America. ISBN 0-8376-0349-8
Audi Part No. LPV 800 422

Audi Fox Service Manual: 1973-1979
Robert Bentley. ISBN 0-8376-0097-9
Audi Part No. LPA 997 082

RB ROBERT BENTLEY, INC. | AUTOMOTIVE PUBLISHERS

Robert Bentley has published service manuals and automobile books since 1950. Please write Robert Bentley, Inc., Publishers, at 1734 Massachusetts Avenue, Cambridge, MA 02138, visit our web site at http://www.rb.com, or call 1-800-423-4595 for a free copy of our complete catalog.

Volkswagen Inspection/Maintenance (I/M) Emission Test Handbook 1980 ▸ 1997

Overview and Troubleshooting for Volkswagen Cars, Vans, and Pickups

|RB|
Robert Bentley
Cambridge, Massachusetts

RB ROBERT BENTLEY, INC. | AUTOMOTIVE PUBLISHERS

Information that makes
the difference.®

1734 Massachusetts Avenue
Cambridge, MA 02138 USA
800-423-4595 / 617-547-4170
http://www.rb.com
e-mail: sales@rb.com

WARNING—Important Safety Notice

In this book we have attempted to describe the operation, service, repair, and modification of Volkswagen fuel injection and emission control systems using examples and instructions that we believe to be accurate. However, the examples, instructions, and other information are intended solely as illustrations and should be used in any particular application only by experienced personnel who are trained in the service, repair and modification of Volkswagen engine control systems and who have independently evaluated the repair, modification or accessory. Implementation of a modification or attachment of an accessory to a Volkswagen engine control system may render the vehicle, attachment, or accessory unsafe for use in certain circumstances.

Do not perform the work described in this book unless you are familiar with basic automotive repair procedures. This book is not a substitute for full and up-to-date information from the vehicle manufacturer or for proper training as an automotive technician. Note that it is not possible for us to anticipate all of the ways or conditions under which vehicles may be serviced or to provide warnings and cautions as to all of the possible hazards that may result.

The vehicle manufacturer referred to in this book will continue to issue service information updates and parts retrofits after the editorial closing of this book. Some of these updates and retrofits may apply to procedures and specifications in this book. We regret that we cannot supply updates to purchasers of this book.

We have endeavored to ensure the accuracy of the information in this book. Please note, however, that considering the vast quantity and the complexity of the service information involved, we cannot warrant the accuracy or completeness of the information contained in this book.

FOR THESE REASONS, NEITHER THE AUTHOR NOR THE PUBLISHER MAKES ANY WARRANTIES, EXPRESS OR IMPLIED, THAT THE EXAMPLES, INSTRUCTIONS OR OTHER INFORMATION IN THIS BOOK ARE FREE OF ERROR, ARE CONSISTENT WITH INDUSTRY STANDARDS, OR THAT THEY WILL MEET THE REQUIREMENTS FOR A PARTICULAR APPLICATION. THE AUTHOR AND PUBLISHER EXPRESSLY DISCLAIM THE IMPLIED WARRANTIES OF MERCHANTABILITY AND OF FITNESS FOR A PARTICULAR PURPOSE, EVEN IF THE AUTHOR OR PUBLISHER HAVE BEEN ADVISED OF A PARTICULAR PURPOSE, AND EVEN IF A PARTICULAR PURPOSE IS INDICATED IN THE BOOK. THE AUTHOR AND PUBLISHER ALSO DISCLAIM ALL LIABILITY FOR DIRECT, INDIRECT, INCIDENTAL OR CONSEQUENTIAL DAMAGES THAT RESULT FROM ANY USE OF THE EXAMPLES, INSTRUCTIONS, OR OTHER INFORMATION IN THIS BOOK. IN NO EVENT SHALL OUR LIABILITY WHETHER IN TORT, CONTRACT OR OTHERWISE EXCEED THE COST OF THIS BOOK.

Your common sense and good judgment are crucial to safe and successful service work. Read procedures through before starting them. Think about how alert you are feeling, and whether the condition of your car, your level of mechanical skill, or your level of reading comprehension might result in or contribute in some way to an occurrence which might cause you injury, damage your car, or result in an unsafe repair. If you have doubts for these or other reasons about your ability to perform safe repair work on your car, have the work done at a qualified shop. **REPAIR OF AUTOMOBILES IS DANGEROUS UNLESS UNDERTAKEN WITH FULL KNOWLEDGE OF THE CONSEQUENCES.**

Before attempting any work on your vehicle, read the warnings and cautions on page viii and any warning or caution that accompanies a procedure in the service book. Review the warnings and cautions on page viii each time you prepare to work on your vehicle.

Copies of this book may be purchased from authorized Volkswagen dealers, from selected booksellers and automotive accessories and parts dealers, or directly from the publisher by mail.

The publisher encourages comments from the readers of this book. These communications have been and will be carefully considered in the preparation of this and other books. Please write to Robert Bentley Inc., Publishers at the address listed on the top of this page.

This book was published by Robert Bentley, Inc., Publishers. Volkswagen has not reviewed and does not vouch for the accuracy of the technical information and procedures described in this book.

Library of Congress Cataloging-in-Publication Data

Volkswagen inspection/maintenance (I/M) emission test handbook,
    1980--1997: overview and troubleshooting for Volkswagen cars, vans,
    and pickups.
       p.   cm.
    Includes index.
    ISBN 0-8376-0394-3 (alk. paper)
    1. Volkswagen automobile---Motors--Fuel injection systems-
-Maintenance and repair--Handbooks, manuals, etc. 2. Volkswagen
automobile--Pollution control devices--Maintenance and repair-
-Handbooks, manuals, etc.
  TL215.V6V6143 1997
  629.25 '3--DC21   97-16347
                                                                         CIP

VWoA Part No. LPV 800 901
Bentley Stock No. VIM7

Editorial closing 01/97

02 01 00 99 98 97   5 4 3 2 1

The paper used in this publication is acid free and meets the requirements of the National Standard for Information Sciences–Permanence of Paper for Printed Library Materials. ∞

Volkswagen Inspection/Maintenance (I/M) Emission Test Handbook 1980–1997—Overview and Troubleshooting for Volkswagen Cars, Vans, and Pickups

©1997 Volkswagen of America, Inc., and Robert Bentley, Inc.

All rights reserved. All information contained in this book is based on the information available to the publisher at the time of first printing. The right is reserved to make changes at any time without notice. No part of this publication may be reproduced, stored in a retrieval system, or transmitted in any form or by any means, electronic, mechanical, photocopying, recording, or otherwise, without the prior written consent of the publisher. This includes text, figures, and tables. This book is published simultaneously in Canada. All rights reserved under Berne and Pan-American Copyright conventions.

Manufactured in the United States of America

Volkswagen Inspection/Maintenance (I/M) Emission Test Handbook 1980 ▸ 1997

Overview and Troubleshooting for Volkswagen Cars, Vans and Pickups

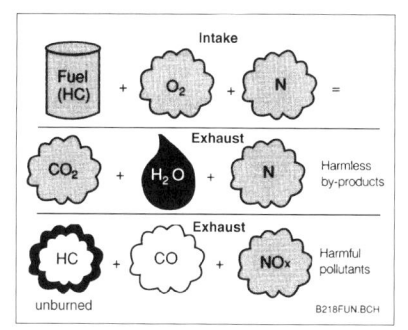

Learn how exhaust emissions are formed and how specific emissions control systems work—Chapter 2

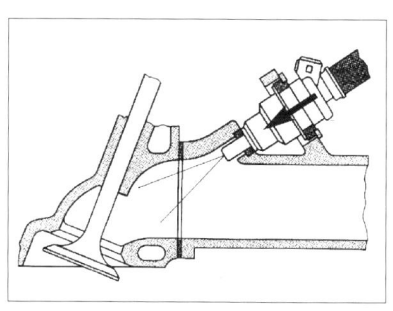

A brief overview of Bosch and Volkswagen fuel injection and emission control systems—Chapter 3

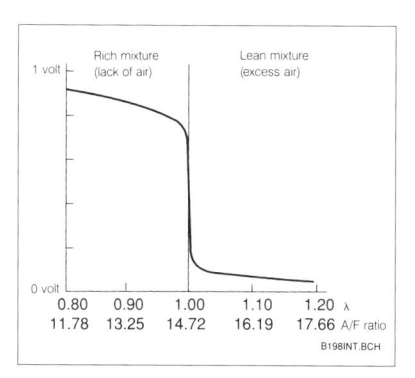

The Lambda control loop is the essence to understanding adaptive controls—Chapter 5

Table of Contents

	Page
Acknowledgement	vi
Foreword	vii
Warnings and Cautions	viii

1 Introduction
How this Handbook is Organized	1
General Information	2
Customer Check List	4
Customer Information about I/M Programs	6

2 Emissions Basics
General	7
The Chemical Reactions of Oxidation and Reduction	9
Combustion By-products	10
The Lambda Value in the Emission Composition	18
Methods to Reduce Emissions	20

3 Engine Management Systems
General	29
Air Flow Control (AFC)	30
Continuous Injection System (CIS)	32
Continuous Injection System–Lambda (CIS-L)	33
Continuous Injection System–Electronic (CIS-E)	34
Continuous Injection System–Electronic, Motronic (CIS-E Motronic)	36
Digijet	38
Digifant	39
Motronic	40

4 Troubleshooting Chart
General	43
Working with the Troubleshooting Chart	43
Evaporative Emissions Control System	46
Checking Dual-Bed Catalytic Converters	49

5 Emission Diagnosis and Repair
General	51
Engine Code AAA	52
Engine Code AAF	55
Engine Code ABA	58
Engine Code ABG	62
Engine Code 9A	65
Engine Code 2H	68
Engine Code CV	71
Engine Code DH	74
Engine Code EJ	77
Engine Code EN	78
Engine Code GX	81
Engine Code HT	84
Engine Code JH	87
Engine Code JN	90
Engine Code KX	93
Engine Code MV	96
Engine Code PF	99
Engine Code PG	102
Engine Code PL	106
Engine Code RD	109
Engine Code RV	112
Engine Code UM	90

Appendix A
Volkswagen Engine Application Chart	116

Acknowledgment

The Publisher wishes to acknowledge that this book is based on a portion of a doctoral thesis by Melanie Werkmeister and Klaus Boetticher, each of Volkswagen AG. While any limitations and errors are entirely the responsibility of the Publisher, this book could never have been published without Ms. Werkmeister and Mr. Boetticher's original research and thesis manuscript.

Foreword

The aim of this handbook is to provide the Volkswagen service technician with information necessary to troubleshoot vehicles that have failed an Inspection and Maintenance (I/M) vehicle emission test. The Environmental Protection Agency's (EPA) latest I/M programs represent a significant increase in sophistication and complexity over the traditional tailpipe or smog test. Clearly, it is more necessary then ever for the technician to carry out a systematic approach when diagnosing I/M failures. The **Volkswagen Inspection/Maintenance (I/M) Emission Test Handbook 1980–1997** is designed to provide the technician with a model-by-model diagnostic tool for I/M test failures.

This handbook is written to support the service aspects of enhanced vehicle emission testing programs, such as the I/M 240 protocol— the most rigid, comprehensive vehicle emission inspection program proposed to date. In its basic form, the I/M 240 test is run on an inertia dynamometer for 240 seconds. The vehicle is operated under varying engine loads while tailpipe exhaust gases are collected and analyzed. If the "cut points" for any of the three exhaust emissions (hydrocarbons, carbon monoxide, and oxides of nitrogen) are exceeded, the vehicle fails the test and requires additional diagnostic work and possible repair.

Although I/M programs are currently in transition, the general pattern of forthcoming programs has begun to emerge. Current and future I/M programs will challenge technicians to improve and maintain their technical skills. Significantly, the EPA has identified the technician as a key ingredient to a successful I/M program: "Skillful diagnostics and capable mechanics are important to assure that failed cars are fixed properly." It is the aim of this handbook to aid the technician in meeting this challenge.

New I/M programs will almost certainly lead to a healthier environment, cleaner running cars, and a business opportunity for the technician who possesses the necessary skills and knowledge to meet the demands. For the Volkswagen service technician, I/M programs will likely be both a challenge and an opportunity for many years to come.

**Robert Bentley,
Publishers**

Please read these warnings and cautions before proceeding with maintenance and repair work.

WARNING—

- Never run the engine unless the work area is well ventilated. Exhaust gas emissions are extremely toxic and can kill.

- Some repairs may be beyond your capability. If you lack the skills, tools and equipment, or a suitable workplace for any procedure described in this book, we suggest you leave such repairs to an authorized dealer service department, or other qualified shop.

- Manufacturers are constantly improving their cars. Sometimes these changes, both in parts and specifications, are made applicable to earlier models. Therefore, before starting any major jobs or repairs to components on which passenger safety may depend, consult your authorized dealer about Technical Bulletins that may have been issued since the editorial closing of this book.

- Do not attempt to work on your car if you do not feel well. You increase the danger of injury to yourself and others if you are tired, upset or have taken medication or any other substance that may keep you from being fully alert.

- Disconnect the battery negative (–) terminal (ground strap) whenever you work on the fuel system or the electrical system. Do not smoke or work near heaters or other fire hazards. Keep an approved fire extinguisher handy.

- Batteries give off explosive hydrogen gas during charging. Keep sparks, lighted matches and open flame away from the top of the battery. If hydrogen gas escaping from the cap vents is ignited, it will ignite gas trapped in the cells and cause the battery to explode.

- Connect and disconnect battery cables, jumper cables or a battery charger only with the ignition switched off, to prevent sparks. Do not disconnect the battery while the engine is running.

- Finger rings, bracelets and other jewelry should be removed so that they cannot cause electrical shorts, get caught in running machinery, or be crushed by heavy parts.

- Tie long hair behind your head. Do not wear a necktie, a scarf, loose clothing, or a necklace when you work near machine tools or running engines. If your hair, clothing, or jewelry were to get caught in the machinery, severe injury could result.

- Illuminate your work area adequately but safely. Use a portable safety light for working inside or under the car. Make sure the bulb is enclosed by a wire cage. The hot filament of an accidentally broken bulb can ignite spilled fuel or oil.

- Do not re-use any fasteners that are worn or deformed in normal use. Many fasteners are designed to be used only once and become unreliable and may fail when used a second time. This includes, but is not limited to, nuts, bolts, washers, self-locking nuts or bolts, circlips and cotter pins. Always replace these fasteners with new parts.

- Never work under a lifted car unless it is solidly supported on stands designed for the purpose. Do not support a car on cinder blocks, hollow tiles or other props that may crumble under continuous load. Never work under a car that is supported solely by a jack. Never work under the car while the engine is running.

- If you are going to work under a car on the ground, make sure that the ground is level. Block the wheels to keep the car from rolling. Disconnect the battery negative (–) terminal (ground strap) to prevent others from starting the car while you are under it.

- Always observe good workshop practices. Wear goggles when you operate machine tools or work with battery acid. Gloves or other protective clothing should be worn whenever the job requires working with harmful substances.

- Catch draining fuel, oil, or brake fluid in suitable containers. Do not use food or beverage containers that might mislead someone into drinking from them. Store flammable fluids away from fire hazards. Wipe up spills at once, but do not store the oily rags, which can ignite and burn spontaneously.

- Do not allow battery charging voltage to exceed 16.5 volts. If the battery begins producing gas or boiling violently, reduce the charging rate. Boosting a sulfated battery at a high charging rate can cause an explosion.

- Some cars covered by this book may be equipped with a supplemental restraint system (SRS) that automatically deploys an airbag in the event of a frontal impact. The airbag is inflated by an explosive device. Handled improperly or without adequate safeguards, it can be accidently activated and cause serious injury.

- Greases, lubricants and other automotive chemicals contain toxic substances, many of which are absorbed directly through the skin. Read manufacturer's instructions and warnings carefully. Use hand and eye protection. Avoid direct skin contact.

CAUTION—

- Manufacturers offer extensive warranties, especially on components of fuel delivery and emission control systems. Therefore, before deciding to repair a car that may still be covered wholly or in part by any warranties issued by the manufacturers, consult your authorized dealer. You may find that he can make the repair for free, or at minimal cost.

- Part numbers listed in this book are for identification purposes only, not for ordering. Always check with your authorized dealer to verify part numbers and availability before beginning service work that may require new parts.

- Before starting a job, make certain that you have all the necessary tools and parts on hand. Read all the instructions thoroughly, do not attempt shortcuts. Use tools appropriate to the work and use only replacement parts meeting manufacturer specifications. Makeshift tools, parts and procedures will not make good repairs.

- Use pneumatic and electric tools only to loosen threaded parts and fasteners. Never use these tools to tighten fasteners, especially on light alloy parts. Always use a torque wrench to tighten fasteners to the tightening torque specification listed.

- Be mindful of the environment and ecology. Before you drain the crankcase, find out the proper way to dispose of the oil. Do not pour oil onto the ground, down a drain, or into a stream, pond or lake. Consult local ordinances that govern the disposal of wastes.

- On cars equipped with anti-theft radios, make sure you know the correct radio activation code before disconnecting the battery or removing the radio. If the wrong code is entered into the radio when power is restored, that radio may lock up and be rendered inoperable, even if the correct code is then entered.

- Connect and disconnect a battery charger only with the battery charger switched off.

- Do not quick-charge the battery (for boost starting) for longer than one minute. Wait at least one minute before boosting the battery a second time.

- Sealed or "maintenance free" batteries should be slow-charged only, at an amperage rate that is approximately 10% of the battery's ampere-hour (Ah) rating.

- Do not allow battery charging voltage to exceed 16.5 volts. If the battery begins producing gas or boiling violently, reduce the charging rate. Boosting a sulfated battery at a high charging rate can cause an explosion.

Chapter 1

Introduction

How this Manual is Organized 1	Customer Check List 4
Inspection and Maintenance (I/M) Programs 2	Customer Information about I/M Programs 6
	Preparation of the Vehicle 6

HOW THIS MANUAL IS ORGANIZED

This handbook is divided into five chapters. The first three chapters provide general information on inspection and maintenance programs, vehicle exhaust emissions, and fuel injection systems. The last two chapters deal specifically with diagnosing and repairing emissions related faults.

Chapter 1—Introduction

This chapter provides a general overview of I/M programs, in particular the I/M 240 program. Also contained in this chapter is a customer check list and a customer information sheet. These forms can be photocopied and completed by the customer or by the Service Advisor in consultation with the customer prior to servicing the vehicle. Properly completed forms can significantly help the technician diagnose emissions related faults.

The customer check list should be reviewed before starting with any I/M troubleshooting, repair, or maintenance. After completion of the work on the car, return the check list to the customer. It contains important information for the customer in case an I/M re-test becomes necessary.

Chapter 2—Emissions Basics

Chapter 2 discusses the combustion process and combustion by-products. It is written in theoretical terms, designed to give the technician a better understanding of various emission control systems and how the systems actually reduce harmful pollutants.

Chapter 3—Engine Management Systems

Chapter 3 describes the fuel injection, ignition, and emission control systems used in 1980 through 1997 Volkswagen cars to which the I/M testing program applies. These systems are briefly described together with a graphic display of the principle functioning components.

Chapter 4—Troubleshooting Chart

Chapter 4 contains a diagnostic chart to help pinpoint emissions related trouble areas. The chart is setup to graphically describe the program that should be used for effective troubleshooting of faults.

This chapter contains general information about the catalytic converter and how the catalytic converter can be tested, including a table on converter efficiency.

Also included is a description of the EVAP-Test, which is currently part of the I/M testing program.

Chapter 5—Emission Diagnosis and Repair

Chapter 5 is organized by engine code and presents engine-specific troubleshooting and diagnostic repair procedures. This chapter is most effective if used in conjunction with the troubleshooting chart found in Chapter 4.

Appendix A— Volkswagen Engine Application Chart

Appendix A lists 1980 through 1997 VW models and their associated engines and engine codes. This chart is helpful when trying to identify a particular engine or fuel system.

Inspection and Maintenance (I/M) Programs

The abbreviation I/M is the Environmental Protection Agency's (EPA) protocol for light car and truck vehicle Inspection and Maintenance programs.

It is the goal of the EPA that all passenger vehicles and light trucks pass a sophisticated emissions inspection. These inspections will likely take place at regular time intervals and be regulated at the state level.

NOTE
- *The information given in this chapter applies mainly to the I/M 240 testing program, although the information is applicable to other EPA I/M vehicle emission testing programs.*

- *Some readers may be familiar with the I/M240 protocol. This was the official emission test for passenger cars and light trucks intended to be introduced in the U.S. on January 1, 1995. The number 240 specifies the duration of the test (240 seconds).*

Within the I/M 240 testing program, the EPA differentiates between geographical areas of low emission problems (basic areas) and geographical areas with severe emission problems (enhanced areas). To accommodate these two areas, two different tests were designed.

In basic areas, only an idle test is performed while in the enhanced area a dynamic test on a dynamometer (roller-test stand) is performed.

The enhanced test is performed under load, meaning that the vehicle is run on an inertia dynamometer according to a specified drive cycle to check the concentrations of hydrocarbons (HC), oxides of nitrogen (NOx) and carbon monoxide (CO). The exhaust gas emissions are sampled at every second during the test.

In addition to the tailpipe emission test, a pressure test is performed to check the fuel tank system for leaks, and a purge test to check the evaporative emission system. A visual check for missing or damaged emission components is also performed.

If a vehicle fails the test in any one of the three categories, it means that the vehicle owner has to repair the fault and have the vehicle re-tested at the emission test station.

The technician who is asked to perform I/M repair will work with a test report issued by the test station. See Fig. 1-1. The technician must understand what repairs or adjustments have to be made based on a review of this report.

I/M 240 EXHAUST EMISSION READINGS

	HC	CO	NOx
State Limits	1.2	20	2.5
Test Reading	0.43	4	0.89
Test Result	Pass	Pass	Pass

EVAP EMISSION SYSTEM TEST

Purge Flow	Pass	Pressure Decay	Pass

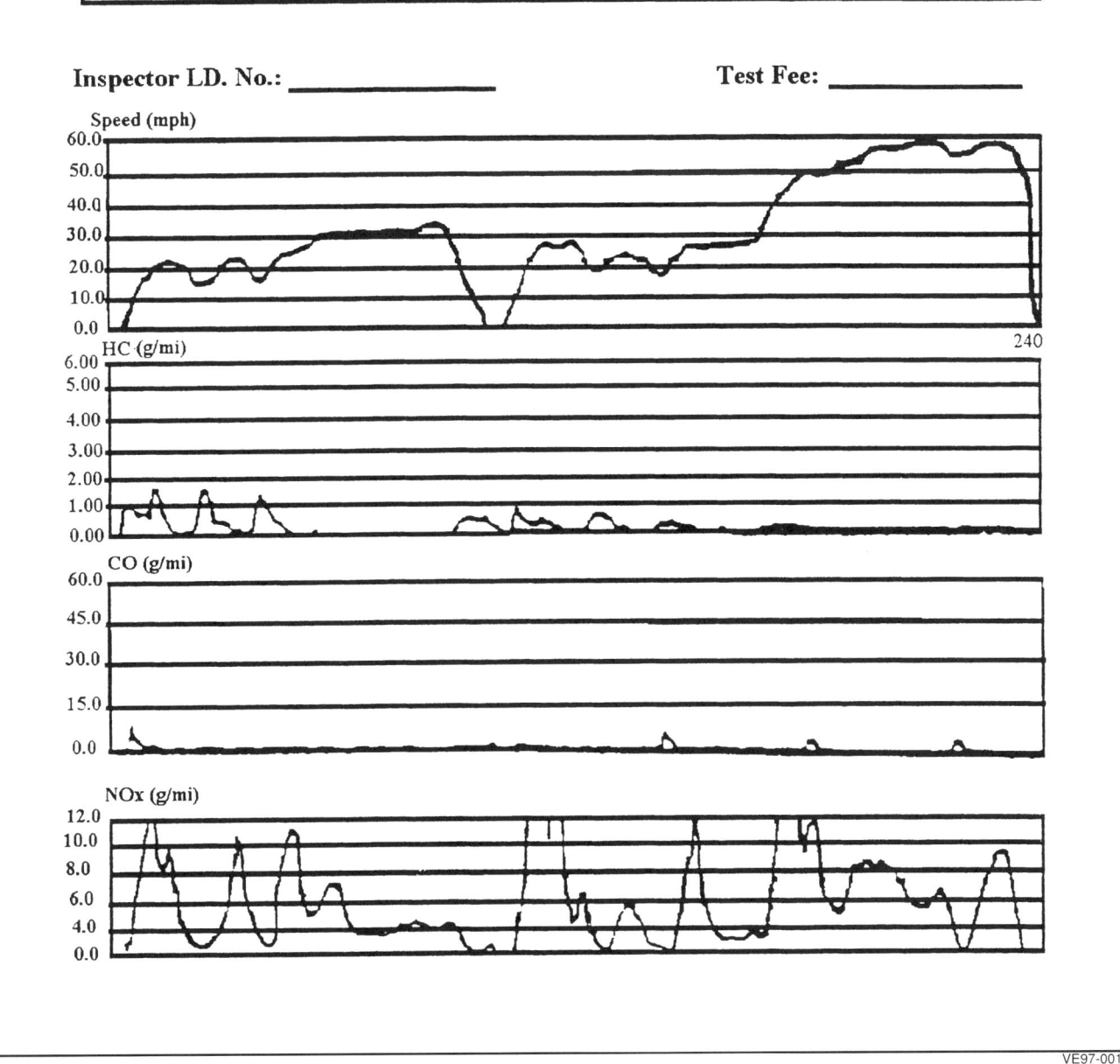

Inspector I.D. No.: _____   Test Fee: _____

Fig. 1-1. Typical Inspection and Maintenance (I/M) test report (I/M 240 report shown).

INSPECTION AND MAINTENANCE (I/M) PROGRAMS

Customer Check List

Below is a typical letter with attached check list that would be provided to the vehicle owner in the event of a failed I/M test. The check list is completed by the vehicle owner and given to the repair technician.

Dear Customer:

Your car did not pass the official Inspection and Maintenance test.

In order to minimize the time your vehicle is in our workshop, we would like you to describe what happened to your vehicle during the test.

Only you can provide us with the information we need for a careful diagnosis of why your vehicle failed the test. Please fill out the form as carefully as possible.

a. Warm-up Phase *(please fill in answers or check off as applicable)*

❏ The vehicle was driven _____ miles to the test station.

❏ You had to wait with your vehicle for _____ minutes before the start of the test.

❏ While in line at the test station did you stop the engine? *yes*___ *no*___

❏ While in line at the test station did you follow the instructions listed below under **Preparation of the Vehicle**. *yes*___ *no*___

❏ Was the vehicle driven by test station personnel for at least two minutes before the test? *yes*___ *no*___

❏ Was the engine fully warmed up? *yes*___ *no*___

❏ Were all electrical consumers (air conditioner, radio, fan, lights, etc.) switched off immediately prior to the test? *yes*___ *no*___

❏ Anything else that we should know about your car?

_____

_____

_____

b. Have you noticed any of the following driveability complaints?

(please fill answers or check off as applicable)

- ❏ Poor engine performance (bucking, poor acceleration) ?   **yes** ___ **no** ___
- ❏ Uneven engine idle?   **yes** ___ **no** ___
- ❏ Engine stops (stalls) when engine is warm?   **yes** ___ **no** ___
- ❏ Engine does not start when cold?   **yes** ___ **no** ___
- ❏ Engine does not start when engine is warm?   **yes** ___ **no** ___
- ❏ Engine continues to run when ignition switch is turned off?   **yes** ___ **no** ___
- ❏ Engine is getting too hot (warning light has come on)?   **yes** ___ **no** ___
- ❏ Engine misfires?   **yes** ___ **no** ___
- ❏ Did the "Check Engine" warning light come on?   **yes** ___ **no** ___
- ❏ Anything else that we should know about your car?

_____

_____

_____

c. Repairs Done *(When were the following repairs done? Please indicate approximate date and the mileage)*

- ❏ Catalytic converter replaced   **Date:** _____ **Mileage:** _____
- ❏ Oxygen sensor replaced   **Date:** _____ **Mileage:** _____
- ❏ Exhaust system welded or repaired   **Date:** _____ **Mileage:** _____
- ❏ Ignition wires or connectors replaced   **Date:** _____ **Mileage:** _____
- ❏ Ignition distributor cap replaced   **Date:** _____ **Mileage:** _____
- ❏ Other repairs done to engine, please describe:

_____

_____

_____

d. Maintenance Work Done *(When were the following maintenance parts replaced? Please indicate approximate date and mileage)*

- ❏ Air cleaner   **Date:** _____ **Mileage:** _____
- ❏ Spark plugs   **Date:** _____ **Mileage:** _____
- ❏ Oil filter   **Date:** _____ **Mileage:** _____
- ❏ Engine oil changed   **Date:** _____ **Mileage:** _____

CUSTOMER INFORMATION ABOUT I/M PROGRAMS

Dear Customer:

Normally a well maintained car will pass an I/M emission test without difficulties. There may be circumstances, however, when the functioning of an emission control system is severely diminished and emission limits are exceeded. For example an old, non-functioning catalytic converter could be the problem. If this is the case, repairs would have to be made and a re-test would be necessary. However, re-tests may be avoided in many instances when a few simple rules are followed:

Preparation of the vehicle

1. Please be sure that the engine is at normal operating temperature when you arrive at the test station. A distance of 5 miles at highway speeds is normally sufficient to reach operating temperature.

2. Should you have to wait in line at a test station, avoid turning the engine off if at all possible.

3. Should you however stop the engine, please observe the following instructions to assure that your engine remains at normal operating temperature:

   - If you have to wait for more than 3 minutes, pull the parking brake and place the shift lever in neutral (N). Start the engine again and raise the engine speed to 1500 to 2000 rpm (watch the tachometer if your car has one). Rev the engine for about a minute.

   > **CAUTION**
   > *Avoid raising the engine speed over 2500 rpm.*

   - Repeat this procedure every 3 to 4 minutes.
   - Shortly before the actual test, switch off all electrical consumers such as air conditioner, radio, fan, lights, etc.

Thank you very much for your help.

Chapter 2
Emissions Basics

General	7
How Emissions are Formed	7
The Chemical Reactions of Oxidation and Reduction	9
Oxidation	9
Reduction	9
Combustion By-products	10
Hydrocarbons (HC)	10
Carbon Monoxide (CO)	12
Oxides of Nitrogen (NOx)	14
Carbon Dioxide (CO_2)	16
Oxygen (O_2)	17
The Lambda Value in the Emission Composition	18
Methods to Reduce Emissions	20
Catalytic Converters	20
Oxidation Catalytic Converters	21
Dual-bed Catalytic Converters	22
Three-way Catalytic Converter	23
Oxygen Sensor and Lambda Regulation	24
Secondary Air Injection	26
Exhaust Gas Recirculation (EGR)	26
Crankcase Ventilation	27
Evaporative Emission (EVAP) Canister System	27

TABLES

a. Fuel/Air Ratio Effects on Engine Operation and Exhaust Emissions ... 19

GENERAL

To operate a car, mostly hydrocarbons (HC) derived from crude oil are used. The chemical energy stored in these compounds is freed during the combustion process and transferred via the crankshaft and transmission to the driven wheels of the vehicle.

Combustion of the air-fuel mixture creates gaseous by-products that make up the exhaust gas. Some of the combustion by-products are harmless, while others have been determined to be harmful. See Fig. 2-1.

How Emissions are Formed

During "complete combustion," hydrocarbons oxygen, and nitrogen produce carbon dioxide, water, and nitrogen. Incomplete combustion can produce carbon monoxide instead of carbon dioxide, hydrocarbons, and oxides of nitrogen. Complete combustion, however, is only a theoretical value. In the real world we can only talk about the actual combustion.

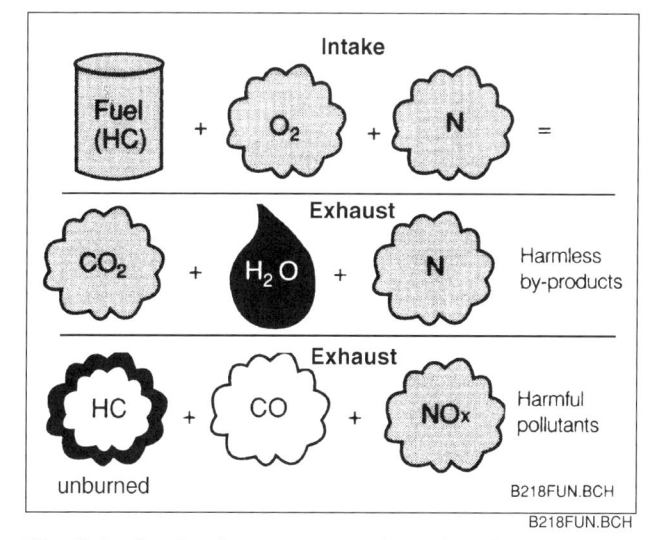

Fig. 2-1. Combustion process produces harmless and harmful by-products.

To discharge as little harmful emissions as possible into the environment and to use the energy in the fuel as much as possible, a precisely adjusted mixture of air and fuel is necessary.

Chapter 2

> **WARNING —**
> Never run the engine unless the work area is well ventilated. Exhaust gas emissions are extremely toxic and can kill.

For a complete combustion we need exactly 1 lb. of fuel and 14.7 lbs. of air. This well known ratio of 14.7:1 is also called the "stoichiometric point" and is necessary for optimum exhaust emissions and engine performance. See Fig. 2-2.

If this ratio of air and fuel is not precisely maintained, the combustion is not optimal and excessive harmful emissions are emitted into the environment. During the actual combustion, the ratio 14.7:1 is not always maintained and results in a greater output of harmful emissions.

To simplify the understanding of this process, the air number Lambda (λ) will be defined as the stoichiometric point of 1.

Lambda also describes the ratio of the air that is actually available for combustion versus the air that is necessary for combustion.

$$\lambda = \frac{\text{air added}}{\text{air needed}}$$

A value of less than (<) 1 identifies rich air/fuel mixture, or not enough air available for complete combustion. This normally results in unburned fuel in the exhaust gas in the form of carbon monoxide (CO) and hydrocarbons (HC).

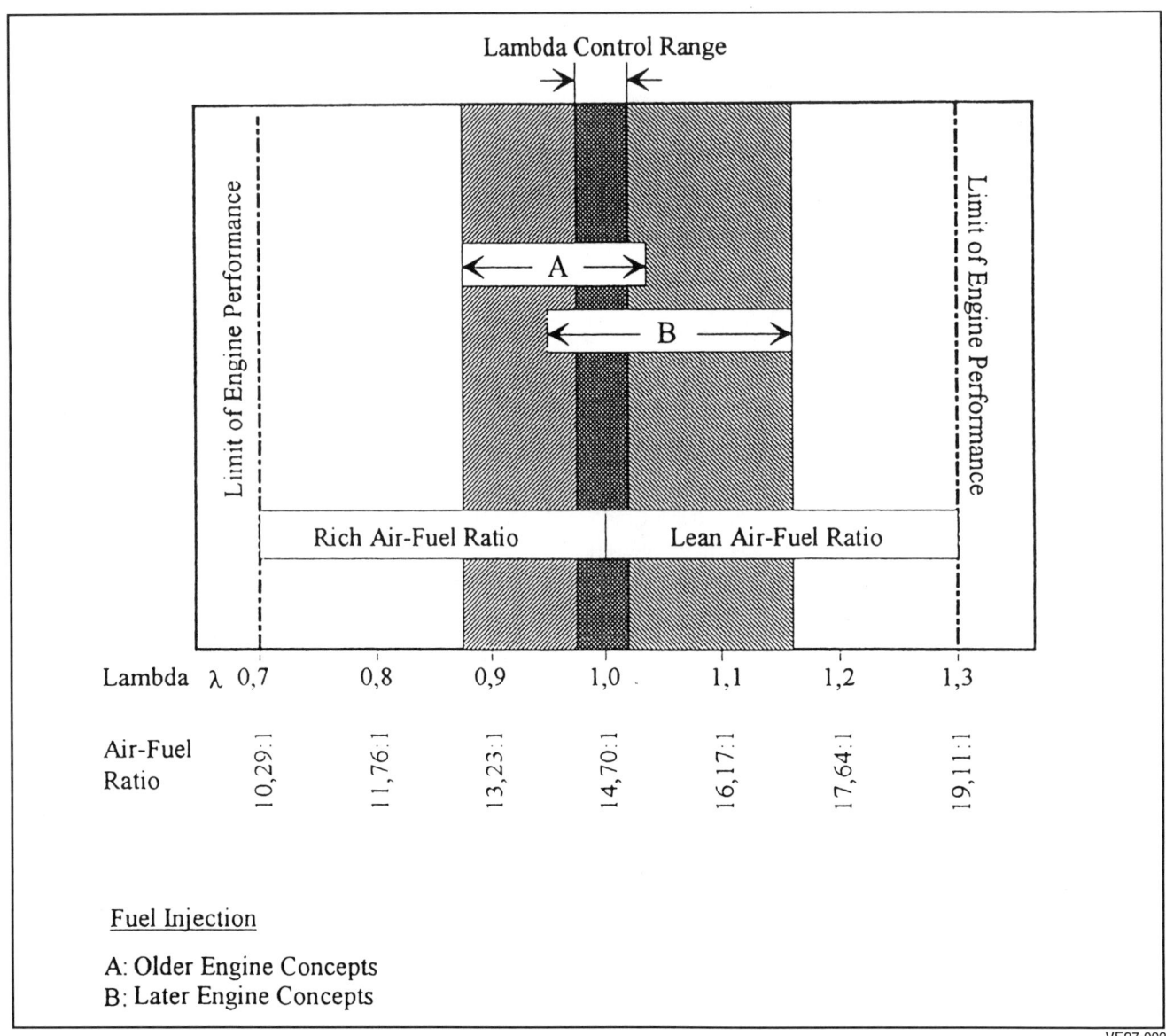

Fig. 2-2. Air value Lambda (λ) control range.

Fuel Injection
A: Older Engine Concepts
B: Later Engine Concepts

Lambda λ	0,7	0,8	0,9	1,0	1,1	1,2	1,3
Air-Fuel Ratio	10,29:1	11,76:1	13,23:1	14,70:1	16,17:1	17,64:1	19,11:1

A value of greater than (>) 1 identifies lean air/fuel mixture, or too much air. In this case there is excess air available for combustion that cannot be used. There is no unburned fuel in the exhaust gas.

THE CHEMICAL REACTIONS OF OXIDATION AND REDUCTION

The chemical reactions that occur during engine combustion are very complex and partially not even known. To understand the relationships during the reduction of noxious emissions it is helpful to know two different chemical reactions that occur in the catalytic converter. The chemical reactions we are talking about are Oxidation and Reduction.

Oxidation

Oxidation means that one or more oxygen atoms are added to a chemical compound. The most important chemical oxidation reactions during the treatment of exhaust gases in the catalytic converter are:

1. Carbon monoxide (CO) oxidizes with oxygen from the air (O_2) to form carbon dioxide (CO_2):

$$2\ CO + O_2 \rightarrow 2\ CO_2$$

2. Hydrocarbons (HC) oxidize with oxygen from the air (O_2) to form water (H_2O):

$$2\ H_2 + O_2 \rightarrow 2\ H_2O$$

Reduction

Reduction means that one or more oxygen atoms are deleted from a chemical compound. The most important chemical reduction reactions during the treatment of exhaust gases in the catalytic converter are:

1. Carbon monoxide (CO) extracts from the oxide of nitrogen (NOx) an oxygen atom. The oxygen atom oxidizes itself to carbon dioxide (CO_2) while the oxide of nitrogen (NOx) is reduced to nitrogen (N_2):

$$2\ NOx + 2\ CO \rightarrow N_2 + 2\ CO_2$$

2. Hydrogen (H_2) extracts from the oxide of nitrogen (NOx) an oxygen atom. The oxygen atom oxidizes itself to water (H_2O) while the oxide of nitrogen (NOx) is reduced to nitrogen (N_2):

$$2\ NOx + 2\ H_2 \rightarrow N_2 + 2\ H_2O$$

COMBUSTION BY-PRODUCTS

> **WARNING —**
> Never run the engine unless the work area is well ventilated. Exhaust gas emissions are extremely toxic and can kill.

The exhaust gas from engine combustion consists mainly of the following components:

- Hydrocarbons **HC** (controlled)
- Carbon Monoxide **CO** (controlled)
- Oxides of Nitrogen **NOx** (controlled)
- Carbon Dioxide **CO_2** (not controlled)
- Oxygen **O_2** (not controlled)
- Water **H_2O** (not controlled)

There are still other components in the exhaust gas. These are, however, omitted to keep this very complex subject simple.

On the following pages, the above mentioned combustion by-products are briefly described.

Hydrocarbons (HC)

> **WARNING —**
> Never run the engine unless the work area is well ventilated. Gasoline hydrocarbons are harmful to your health.

Gasoline is a hydrocarbon mixture. The general description of hydrocarbons is a collective name for organic compounds that consist only of carbon (C) and hydrogen (H) molecules.

If hydrocarbons are detected during an emissions test, unburned or not completely burned fuel is being exhausted from the combustion chamber. Also engine oil—although not easily burned—can get into the combustion chamber and cause high HC readings.

Ignition problems (worn/fouled spark plugs, corroded plug connectors, faulty wiring, incorrectly adjusted ignition, etc.) are the most common cause of high hydrocarbons readings. Another common cause is due to lean fuel/air mixtures. In this case, the fuel/air mixture is so lean that it is hard to ignite and burn. The lean mixture can be caused by poor mixture, but more likely it is due to intake air leaks (leaking vacuum hoses or connections). See Fig. 2-3.

The HC emission can also be caused by the so-called "quenching effect" which is a design problem of the combustion chamber. Due to uneven heat distribution, the cylinder charge is not fully burned and this unburned fuel is emitted into the exhaust system.

A reduction of the HC-emission can be chemically produced by oxidation (introducing oxygen into the exhaust gas). This is achieved through one of two ways; either by using a Three-Way Catalytic Converter (TWC) or through Secondary Air Injection (AIR). The TWC has the ability to store oxygen on its large surface. The AIR system pumps air into the exhaust and works together with an oxidation-type catalytic converter.

> *Health effects—*
>
> *Hydrocarbons irritate the mucous membrane. They can be identified as follows:*
>
> *Saturated compounds (paraffin) are nearly odorless and have a narcotic effect. Unsaturated compounds (olefin, acetylene) have a slightly sweet smell and together with NO are to a great extent responsible for smog.*
>
> *The aromatic hydrocarbons have a characteristic, often pleasant smell. They however have a harmful effect on the nervous system and may also produce cancer.*

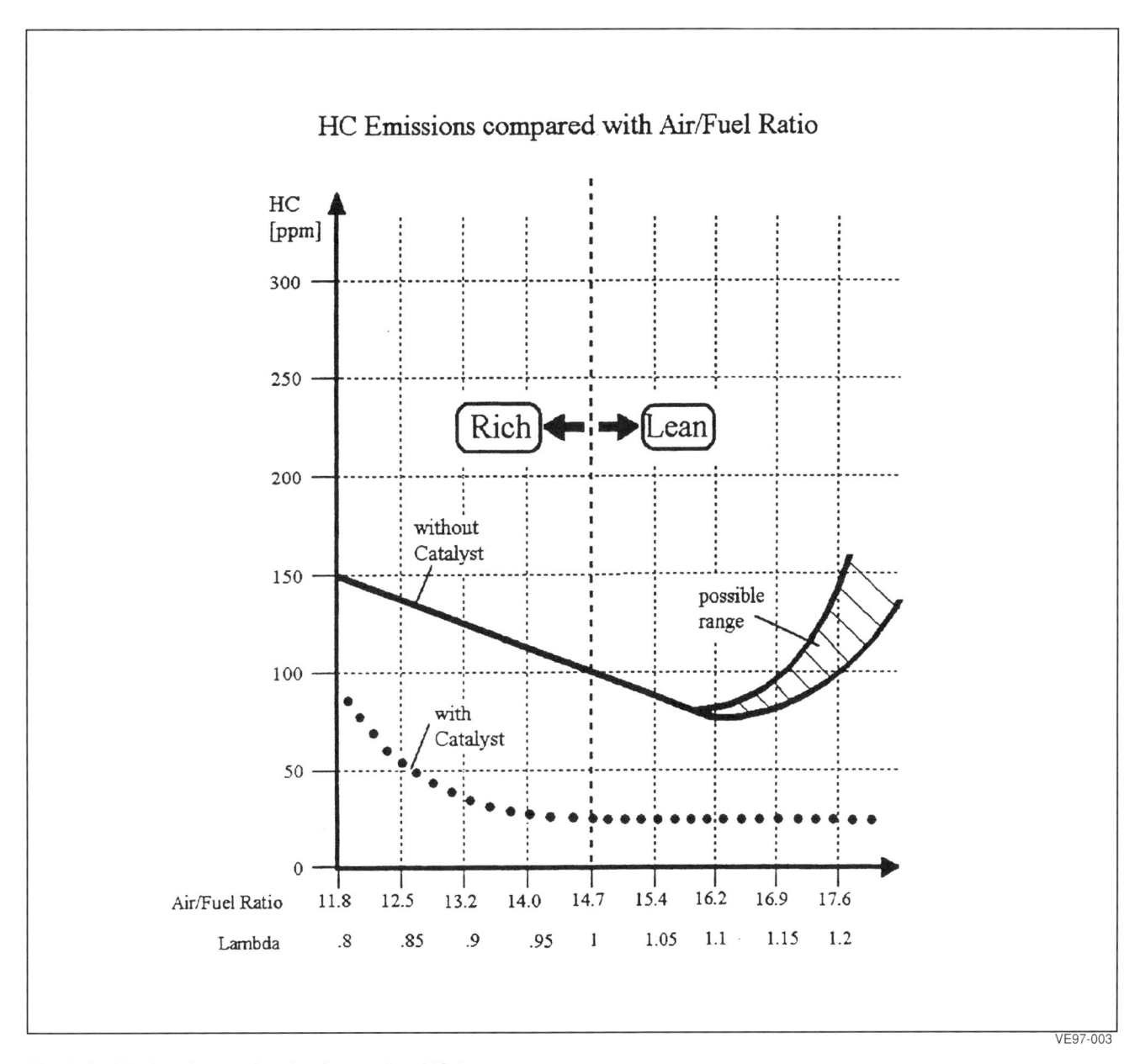

Fig. 2-3. Tracing the combustion by-product HC in connection with Lambda (λ).

Carbon Monoxide (CO)

> **WARNING —**
> - Never run the engine unless the work area is well ventilated. Carbon monoxide is very toxic and can kill.
> - Carbon monoxide is easily and quickly absorbed into the bloodstream and takes a long time to leave the bloodstream.

High CO-values are associated mostly with a mixture that is too rich, or in other words ($\lambda < 1$). This means there is not enough oxygen available for the combustion of fuel. Anything that prevents the supply of fresh air to the combustion chamber can increase the CO emission. We therefore can use the CO content of the exhaust gas as a direct indicator for the air/fuel ratio of the mixture. See Fig. 2-4.

A reduction of the CO-emission can also be chemically produced by oxidation (introducing oxygen into the exhaust gas). This is achieved through a Three-Way Catalytic Converter (TWC) or through a Secondary Air Injection (AIR) system.

> *Health effects—*
>
> *CO is a very toxic, colorless, tasteless, and odorless gas that combines easily with the blood molecule hemoglobin. This effectively replaces the oxygen in the blood and can quickly lead to toxic poisoning and asphyxiation.*

Emissions Basics 13

CO Emissions compared with Air/Fuel Ratio

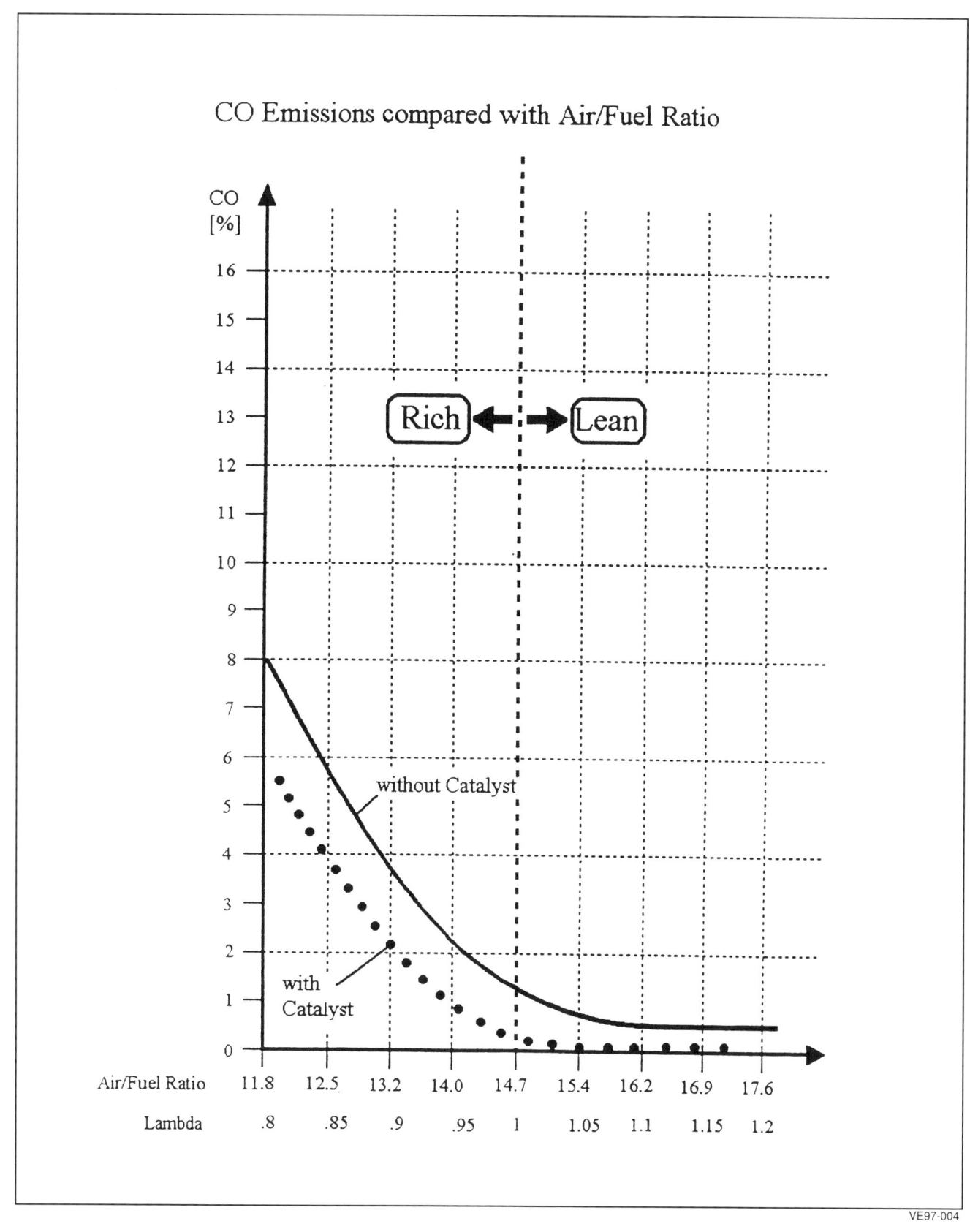

Fig. 2-4. Tracing the combustion by-product CO in connection with Lambda (λ).

COMBUSTION BY-PRODUCTS

Oxides of Nitrogen (NOx)

WARNING —
Never run the engine unless the work area is well ventilated. Oxides of nitrogen are harmful to your health.

Oxides of nitrogen (NOx) appear in the form of nitrogen monoxide (NO) and nitrogen dioxide (NO_2) and are also a collective name for many compounds that need not be identified.

NOx are generated during the combustion of nitrogen dioxide (NO_2) under very high temperatures and pressures. Normally, nitrogen (composes 79% of air) is a low reaction gas that combines with the oxygen from the air to form oxides. See Fig. 2-5.

High combustion temperatures are created by high compression ratios, very lean fuel/air mixtures and early ignition timing.

NOx emissions are mainly controlled by measures that lower the combustion chamber temperature. This can be done by lowering compression ratios, adjusting ignition timing toward late, applying exhaust gas recirculation, and through control of fuel/air ratios.

Health effects—

NO is a colorless gas that replaces the oxygen of the blood molecule hemoglobin, which can quickly lead to symptoms of paralysis.

NO2 is a reddish-brown gas that has a stinging smell which can lead to tissue damage in the lungs.

In connection with non-saturated hydrocarbons and ultraviolet sun radiation, NOx emissions are main contributors for smog. Because of this reason they are to be considered noxious emissions.

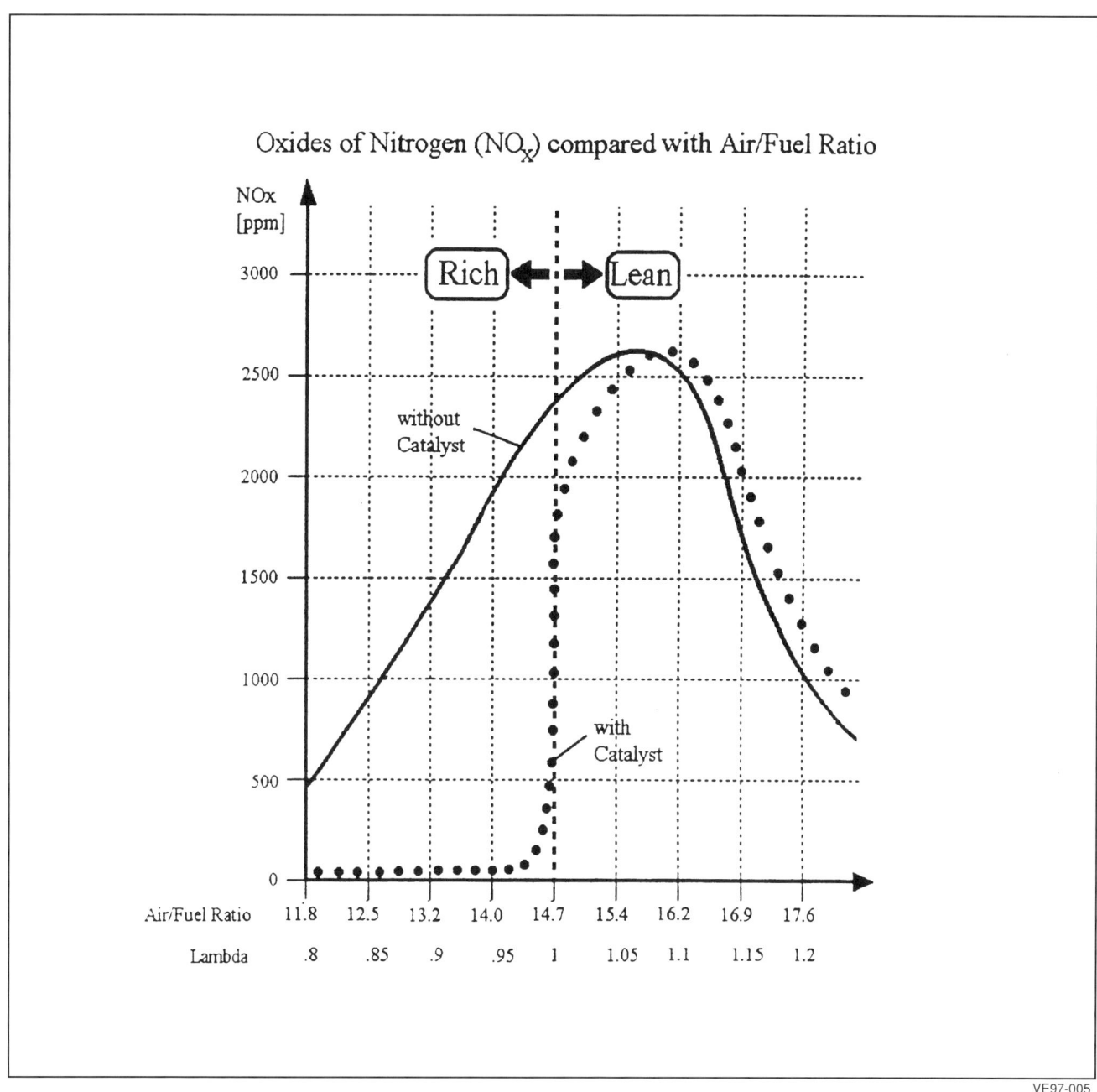

Fig. 2-5. Tracing the combustion by-product NOx in connection with Lambda (λ).

Chapter 2

Carbon Dioxide (CO₂)

The value for carbon dioxide is not limited. High values (14% to 15% in the exhaust gas) are reached when $\lambda = 1$ (stoichiometric point). This is an indication that the fuel/air mixture is properly adjusted. See Fig. 2-6.

> **Health effects—**
>
> *CO2 in normal concentration is not health threatening, however it destroys the ozone layer of the earth and is therefore considered to be harmful to the earth's atmosphere.*

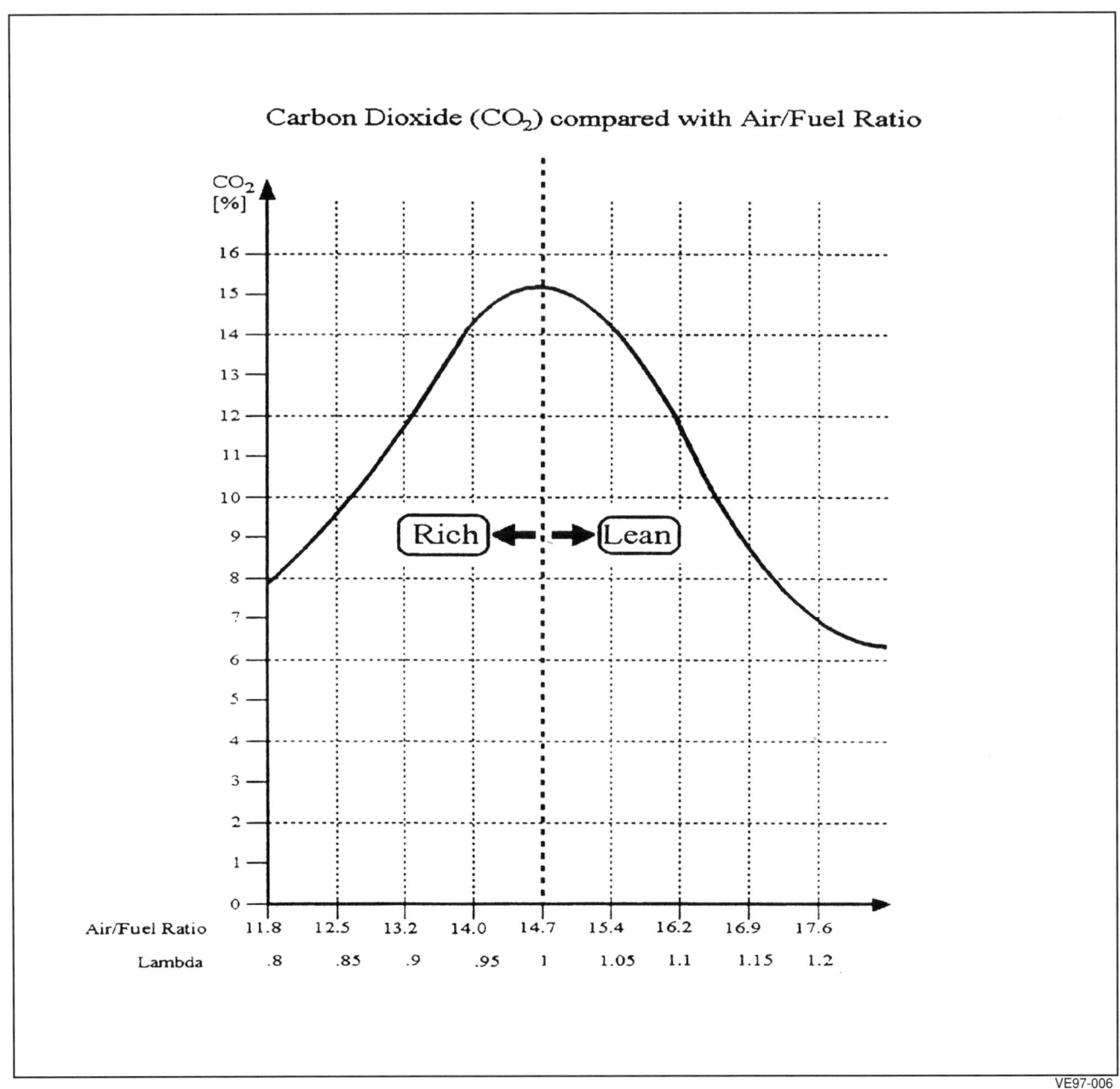

Carbon Dioxide (CO₂) compared with Air/Fuel Ratio

Air/Fuel Ratio	11.8	12.5	13.2	14.0	14.7	15.4	16.2	16.9	17.6
Lambda	.8	.85	.9	.95	1	1.05	1.1	1.15	1.2

Fig. 2-6. Tracing the combustion by-product CO₂ in connection with Lambda (λ).

COMBUSTION BY-PRODUCTS

Oxygen (O$_2$)

Air (which contains up to 21% oxygen) is mixed with fuel for combustion in the combustion chamber. The oxygen content measured in the exhaust gas makes a lean mixture when the oxygen is high and a rich mixture when the oxygen content is low.

Carbon monoxide (CO) behaves exactly opposite to that of oxygen. In other words, a high CO content indicates a rich mixture, and a low content indicates a lean mixture.

At the stoichiometric point ($\lambda = 1$), the amount of carbon monoxide and oxygen are about evenly low. See Fig. 2-7.

> *Health effects—*
>
> ***Oxygen is not considered to be a harmful emission. It is more like an information gas—that by its concentration you can tell whether the exhaust gas is rich or lean.***

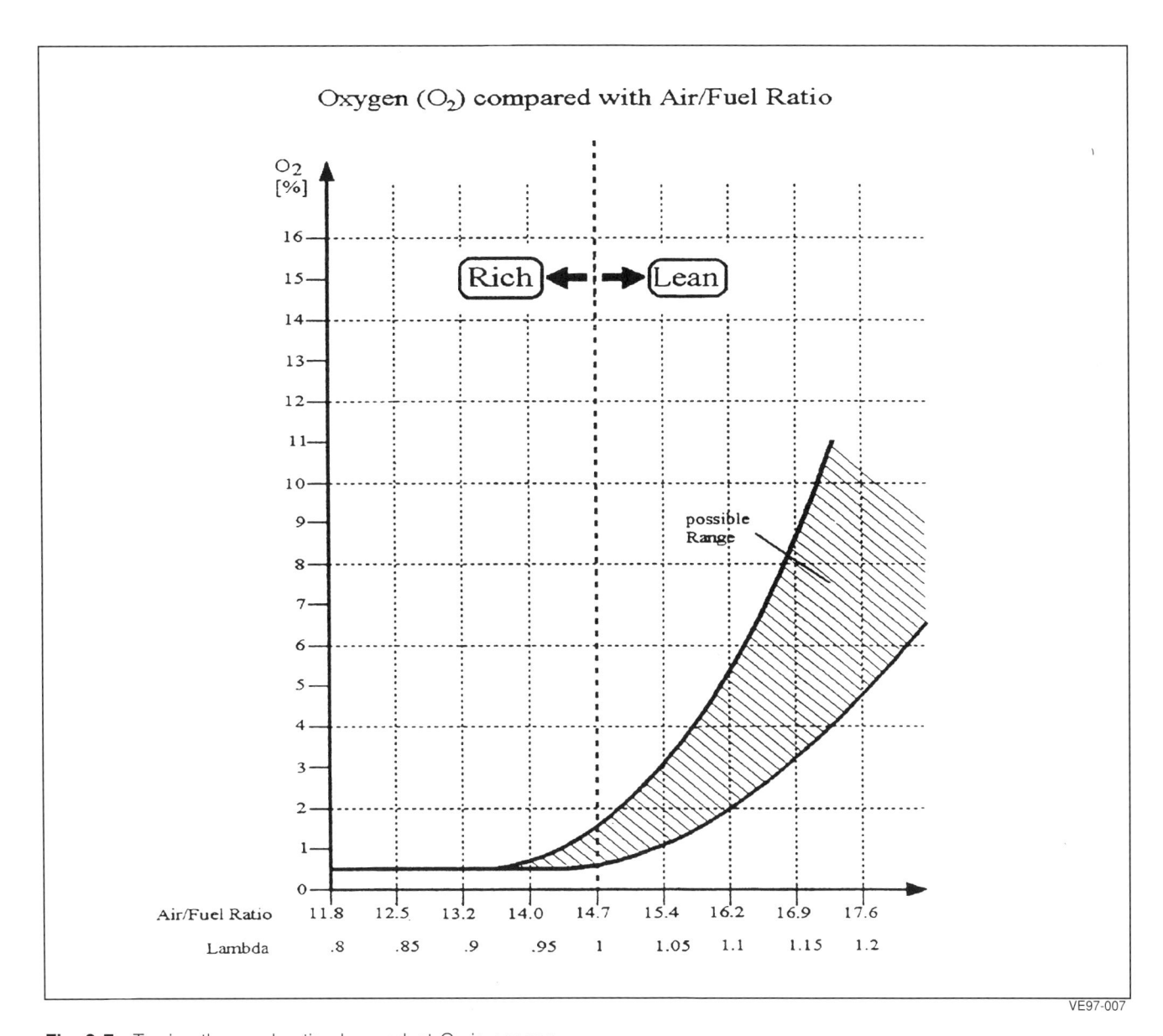

Fig. 2-7. Tracing the combustion by-product O$_2$ in connection with Lambda (λ).

COMBUSTION BY-PRODUCTS

THE LAMBDA VALUE IN THE EMISSION COMPOSITION

For the best combustion process, a precisely adjusted fuel/air ratio is very important. At a fuel/air ratio of 14.7:1, $\lambda=1$. This point is also called the "stoichiometric point" and indicates the optimum exhaust and engine operating conditions. See Fig. 2-8.

The narrow range of 0.97 to 1.03 is called the "Lambda window" because of its good exhaust and engine operation conditions. In this range a Three-Way Catalytic Converter must be used. The Lambda window can be recognized by the following:

- Low carbon monoxide (CO) and oxygen (O_2) content (0.2 to 1.5%)
- High carbon dioxide (CO_2) (10 to 15%) and oxides of nitrogen (NOx) (2000 to 2500 PPM)
- Low hydrocarbons (HC) (< 200 PPM)

During engine combustion this "stoichiometric" condition varies to a certain degree because of the changing load and other operating conditions. In this way the fuel/air ratio changes away from the stoichiometric point toward the rich or the lean range and changes the composition of the exhaust gas.

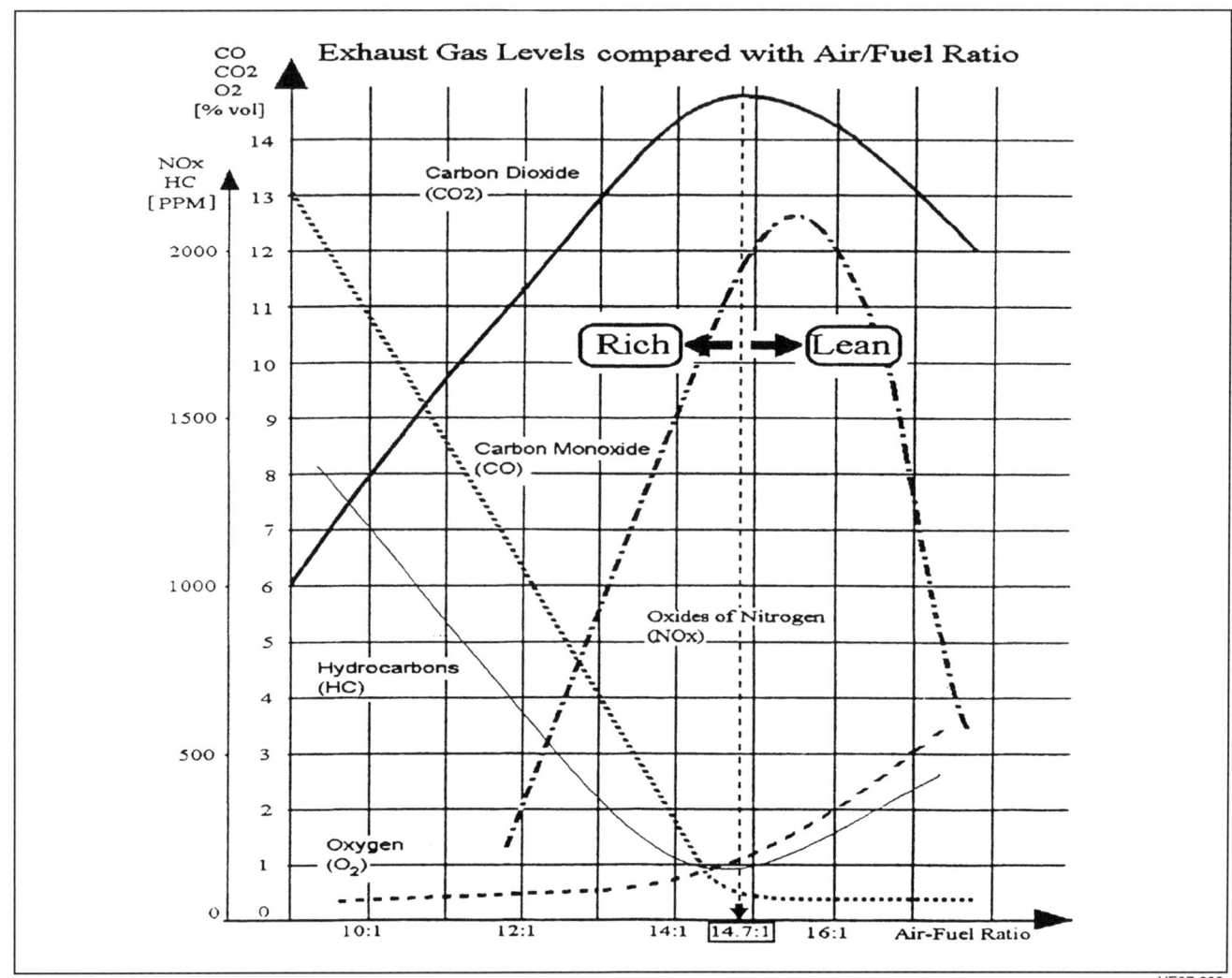

Fig. 2-8. Influence of the air/fuel ratio on exhaust gas.

The following table provides a brief overview of how the fuel/air ratio affects engine operation and exhaust emissions.

Table a. Fuel /Air Ratio Effects on Engine Operation and Exhaust Emissions

Fuel /Air Ratio	Effect
Too rich (Lambda (λ) window = 0.8 to 0.9)	• high CO-value • high HC-values • black exhaust gas • engine oil thinning • high fuel consumption
Enriched range	• high CO-value • high HC-values • enhanced engine performance • increased fuel consumption • less tendency to knock
Stoichiometry (λ=1, approx.) (Lambda (λ) window = 0.97 to 1.03	• optimal idle • good emission values
Slightly lean range	• low HC-values • low CO-values • maximum NOx-values • reduced engine performance • low fuel consumption • slight tendency to knock
Lean range (Lambda (λ) window = 1.1 to 1.2)	• increased NOx-values • poor engine performance • misfiring • temperature damage to pistons and valves • high tendency to knock

THE LAMBDA VALUE IN THE EMISSION COMPOSITION

METHODS TO REDUCE EMISSIONS

To reduce harmful emissions the following measures are used:

- Catalytic converters
- Lambda regulation
- Secondary air injection (AIR)
- Exhaust gas recirculation (EGR)
- Crankcase ventilation
- Evaporative canister system (reduction of fuel evaporation)

Catalytic converters and secondary air injection are part of the exhaust gas treatment. This includes all measures that allow additional chemical change of the exhaust gas composition.

In addition to systems that treat the exhaust gas itself, engine design directly affects the composition of the exhaust gas. Primarily, the various types of fuel injection and ignition systems as well as exhaust gas recirculation, crankcase ventilation and the EVAP canister systems.

Catalytic Converters

The catalytic converter is a metal housing filled with a carrier material. The carrier material is normally a ceramic monolith (AL_2O_3) covered by an alloy of precious metals. Precious metals have a very slow chemical reaction without taking part in the reaction, but stimulate chemical reactions in other combustion by-products.

The job of the catalytic converter is to react to the pollutants in the exhaust gas (HC, CO and NOx) and convert them into the non-harmful emissions (H_2O, CO_2 and N_2).

Although the catalytic converter promotes chemical reactions without affecting the catalyst itself, the catalytic converter will age over time due to high temperatures. The converter can also be damaged by knocks to the housing and the use of leaded gasoline.

Emissions Basics 21

For the catalytic converter to operate properly, it must be at sufficient operating temperature. Below 300°C (572°F) the conversion does not take place.

The following types of catalytic converters are used:

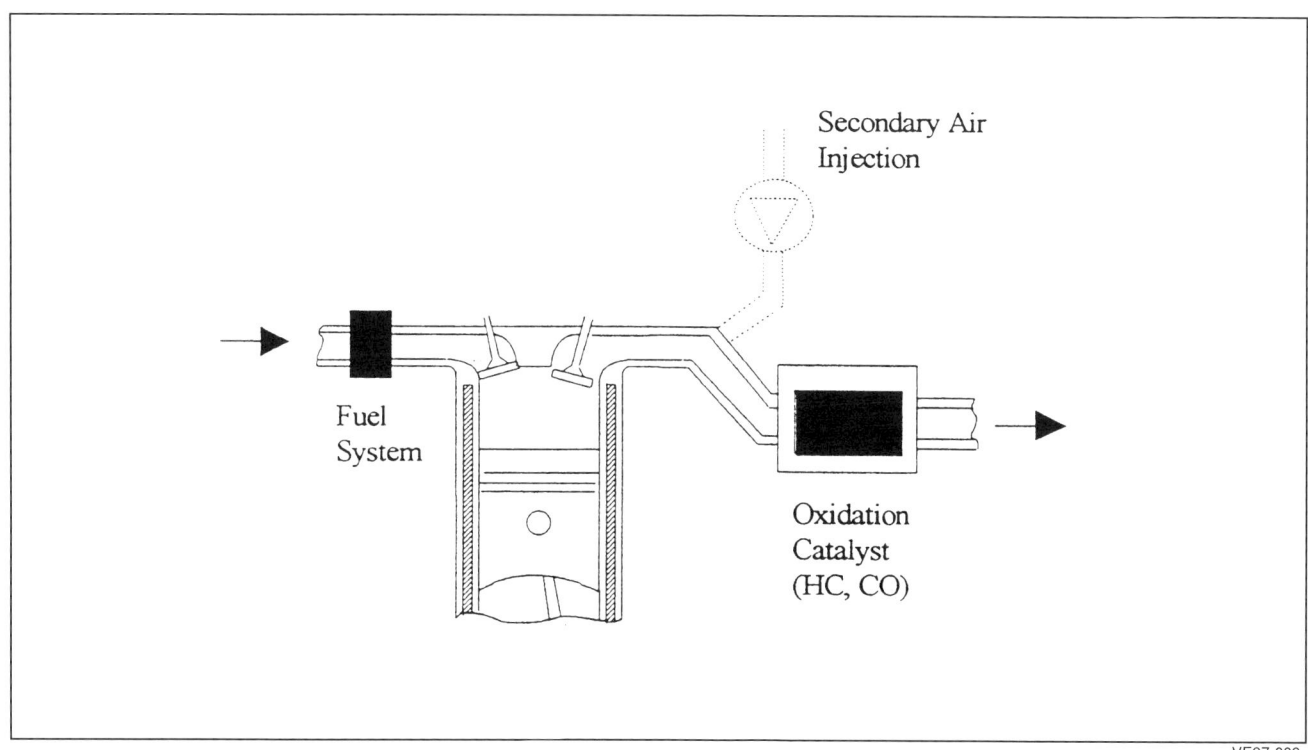

- Oxidation catalytic converters
- Dual-bed catalytic converters
- Three-way catalytic converters

In addition there are systems that do not affect the oxygen content in the exhaust gas and where the catalytic converter works alternately as a reduction or oxidation converter.

Oxidation Catalytic Converters

Oxidation catalytic converters can only convert the harmful emissions HC and CO into H_2O and CO_2. The conversion requires a lot of oxygen supplied primarily as secondary air. See Fig. 2-9.

If the engine runs with a lean mixture ($\lambda > 1$), not all of the oxygen is used for the combustion. This effectively leaves sufficient oxygen for the after-treatment of the exhaust gas.

Fig. 2-9. Oxidation catalytic converter.

METHODS TO REDUCE EMISSIONS

Dual-bed Catalytic Converters

Dual-bed catalytic converters consist of two converters, one positioned behind the other in the same housing. See Fig. 2-10.

In the first converter, oxides of nitrogen (NOx) are converted in the exhaust gas. Normally, this reaction can only happen when there is little or no oxygen in the exhaust gas (i.e. a rich fuel mixture). By employing a reduction converter, the result is proper fuel mixtures and low (normal) fuel consumption.

In the second converter, CO and HC are oxidized to H_2O and CO_2. Normally, this reaction can only take place if there is sufficient oxygen in the exhaust gas. For this, additional air is fed between the two converters by means of secondary air injection (see also **Secondary Air Injection**).

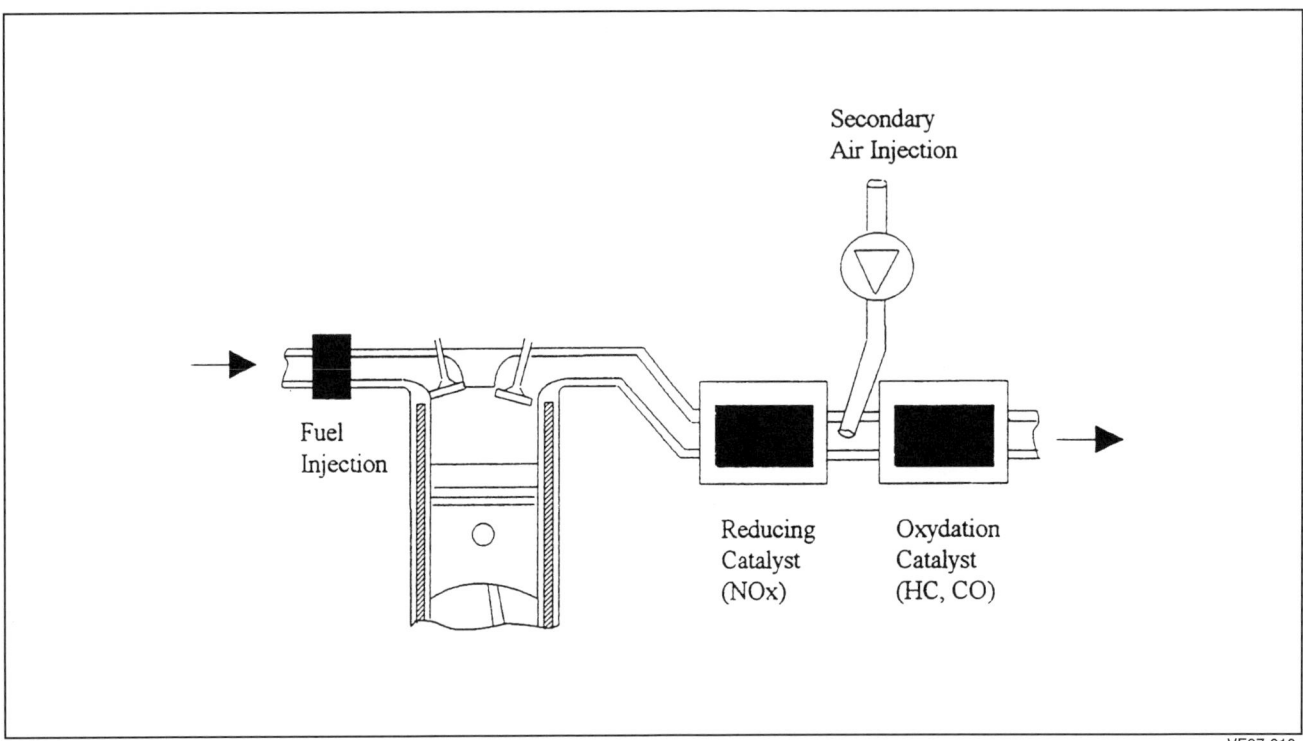

Fig. 2-10. Dual-bed catalytic converter.

METHODS TO REDUCE EMISSIONS

Three-way Catalytic Converter

The three-way catalytic converter combines the principles of the reduction and oxidation converters. The three harmful emissions HC, CO and NOx are converted in just one converter.

To accomplish this, oxygen is stored by the converter when the exhaust gas is rich in oxygen. This stored oxygen is released as necessary for the oxidation and conversion to H_2O and CO_2 while NOx is reduced.

When the engine runs lean, CO and HC become H_2O and CO_2, however NOx is chemically not converted and passes the converter unchanged.

To function as described, the engine must run with a precisely adjusted mixture of λ = 0.97 to 1.03. This narrow range is called the "Lambda window." This way, harmful emissions are reduced up to 90%.

To make it possible for an engine to run under almost all load conditions at λ =1, a Lambda regulation system is needed. With the aid of an oxygen sensor located in the exhaust system ahead of the catalytic converter it can be determined whether there is residual oxygen in the exhaust gas.

The oxygen sensor reports back to the electronic engine control module (ECM), which in turn evaluates the information and initiates a corrective measure. When the exhaust gas is rich, the mixture is adjusted toward lean and when lean the mixture is adjusted toward rich. See **Oxygen Sensor and Lambda Regulation** given below for more information

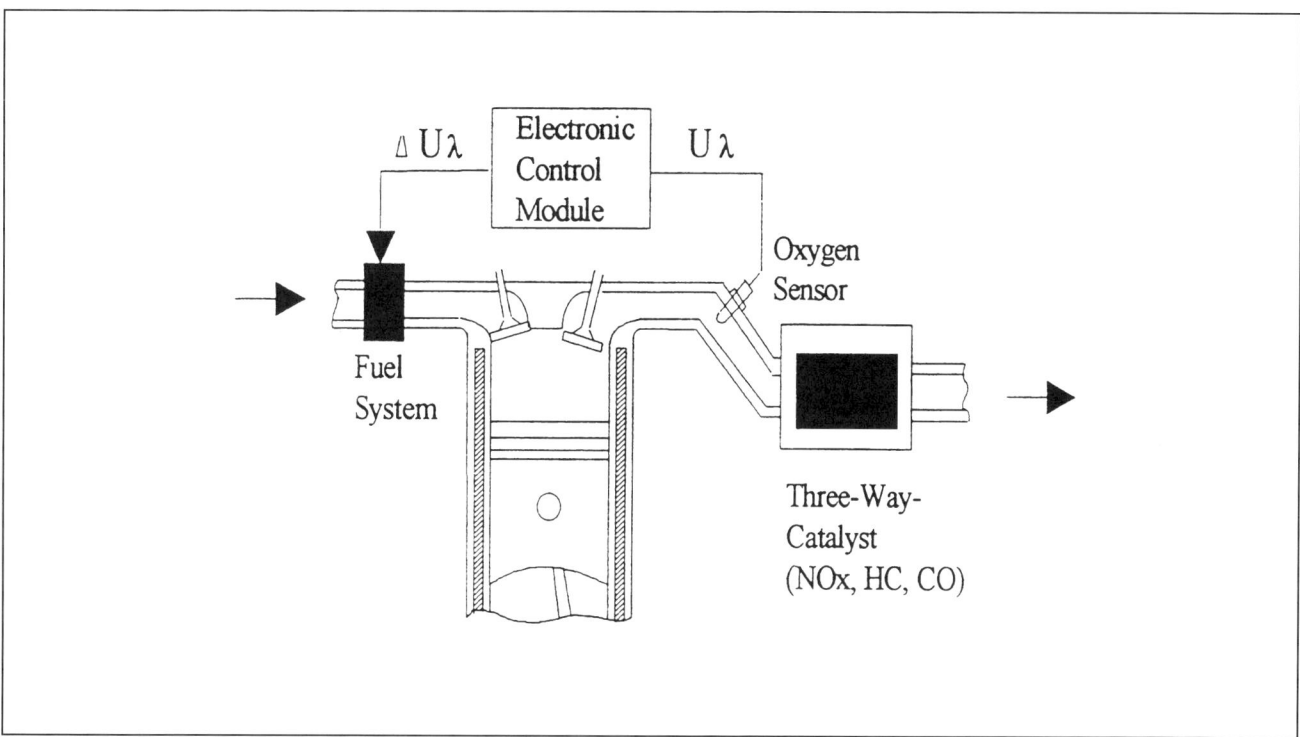

Fig. 2-11. Three-Way catalytic (TWC) converter.

METHODS TO REDUCE EMISSIONS

Oxygen Sensor and Lambda Regulation

Precise regulation of the air-fuel mixture can only be accomplished through a reporting procedure regarding the composition of the exhaust gas. This assignment is accomplished by the Lambda regulation system, which operates in a closed control loop. This control loop constantly measures the residual oxygen content in the exhaust gas. The volume of fuel needed is then immediately corrected for optimum combustion. See Fig. 2-12.

The oxygen sensor acts as the measuring sensor that creates a current differential at $\lambda = 1$. This current differential allows a judgment whether the exhaust gas is richer or leaner than $\lambda = 1$.

The oxygen sensor is installed into an appropriate location in the exhaust system with its ceramic part protruding into the exhaust gas and with the other part exposed to the outside air. The surface of the ceramic part is permeable to oxygen at approximately 300°C (572°F) and becomes electrically conductive.

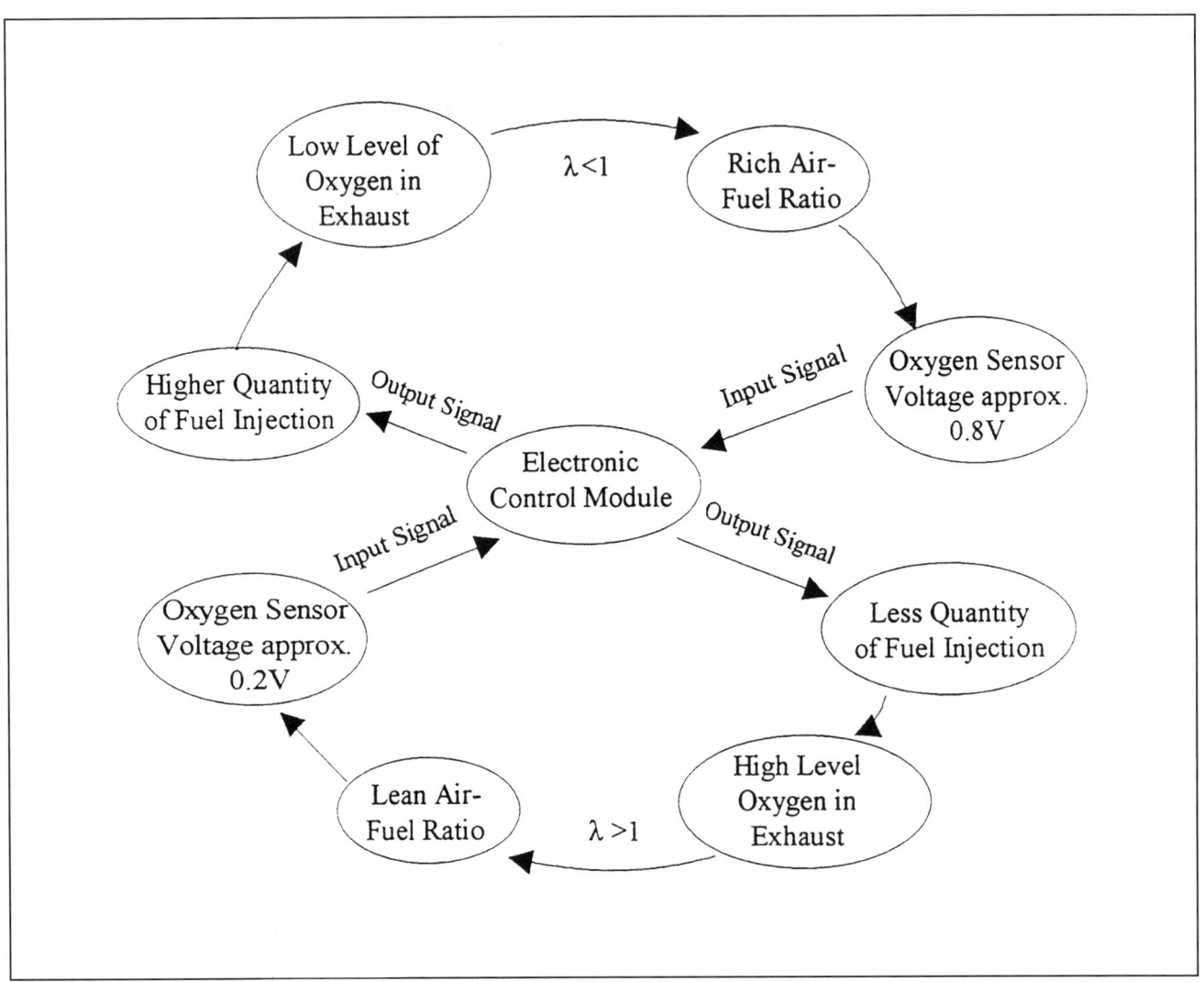

Fig. 2-12. Current path of the oxygen sensor signal.

METHODS TO REDUCE EMISSIONS

Emissions Basics 25

If the oxygen content outside the sensor is different from the oxygen content in the exhaust system an electrical current is produced. This current is used to measure the difference of the oxygen content on both side of the sensor. The engine control module evaluates the signal and initiates the correction. See Fig. 2-13.

The oxygen sensor requires an operating temperature of at least 300°C (572°F) and is therefore also supplied as a heated oxygen (three-wire or four wire) sensor for faster response after a cold start.

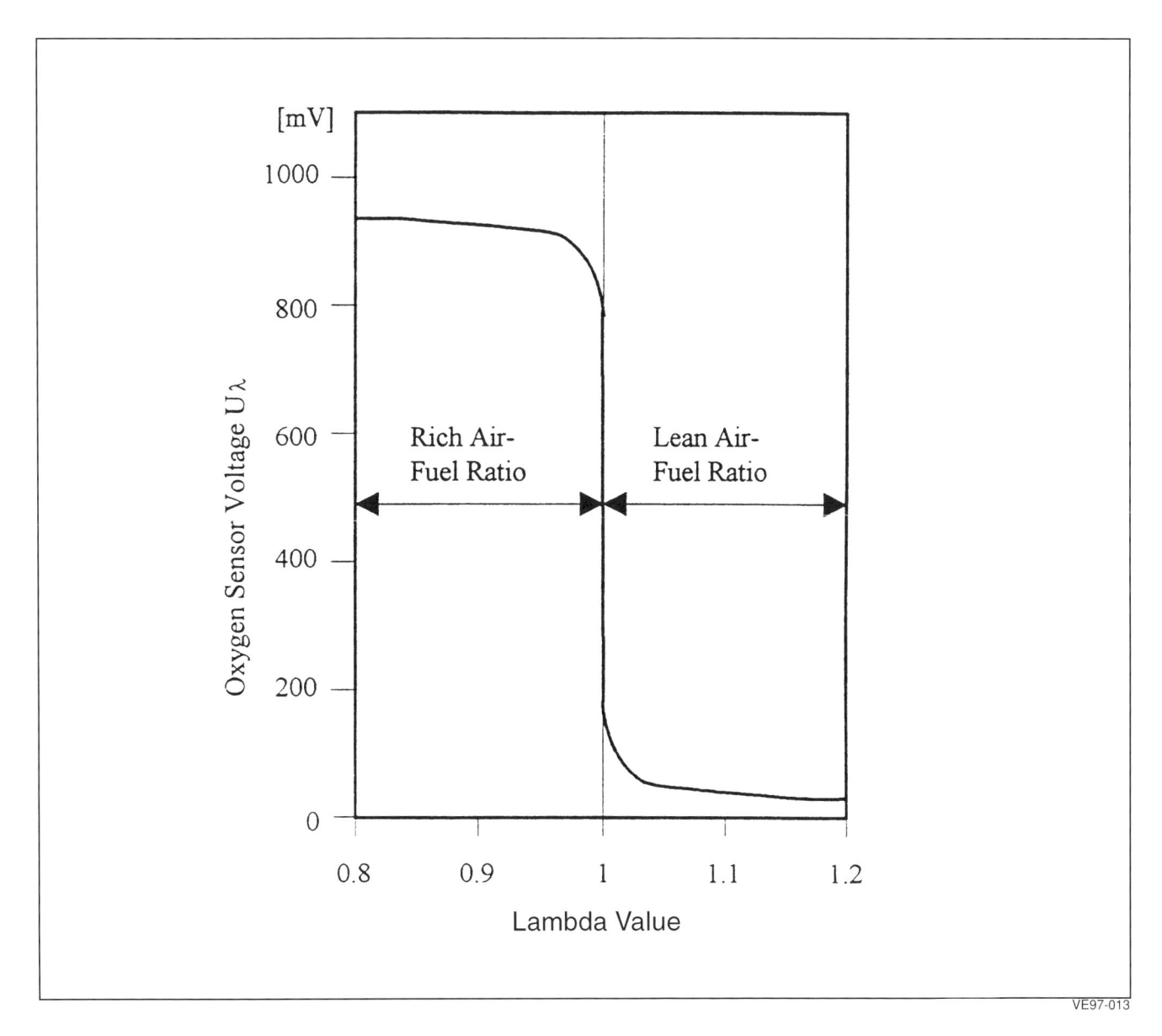

Fig. 2-13. Current differential of Lambda (λ) signal. Note the sensitivity of the Lambda sensor at λ=1. With a voltmeter connected to sensor signal wire and the engine fully warmed up, sensor voltage should fluctuate rapidly up and down around 450-500mV, ranging from about 200 mV to about 800 mV.

METHODS TO REDUCE EMISSIONS

Secondary Air Injection

By injecting additional air into the hot exhaust gas in the exhaust system, it makes after-burning of unburned hydrocarbons (HC) and carbon monoxide (CO) possible and their conversion to water (H_2O) and carbon dioxide (CO_2).

For systems with oxidation catalytic converters, the fresh air is injected in front of the converter to supply sufficient oxygen for the oxidation of HC and CO (see also **Oxidation Catalytic Converter**). The control of the air supply is done today by an electrically-controlled solenoid valve.

In some systems, secondary air injection is used to shorten the heat-up phase of the catalytic converter after a cold start. After a cold start there is a high proportion of unburned fuel in the exhaust gas. The secondary air causes an after-burning of the fuel which brings the converter quickly to operating temperature.

Exhaust Gas Recirculation (EGR)

In exhaust gas recirculation, a portion of the exhaust gas is diverted via a connecting line into the air intake. An EGR valve in the line controls the amount of re-circulated exhaust gas. On later cars, a temperature sensor also measures the temperature of the recirculated gas which can be used as an additional control value.

The recirculated exhaust gas causes a direct reduction of the oxygen content in the fuel/air mixture and a reduction of the temperature in the combustion chamber because of the high temperature absorption of the exhaust gas. This causes the combustion temperature to be lowered and with that a reduction of NOx emission by up to 60%.

Too much recirculated exhaust gas causes an increase of HC and CO emissions and affects the combustion negatively to the point of misfiring and poor running. The amount of recirculated exhaust gas for a 4-cycle engine is therefore limited to approximately 3-5%.

Crankcase Ventilation

Combustion gases escaping between pistons and cylinders (blow-by gases) enter the crankcase and cause a pressure increase as well as a thinning of the engine oil. To prevent this from happening, the crankcase must be vented.

In older systems, the blow-by gases together with the oil vapors were vented to the outside air which lead to higher HC entering the atmosphere. In newer systems the gases from the crankcase are fed back into the intake system and are burned again.

A crankcase ventilation valve opens to the intake system when there is slight pressure in the crankcase. When the engine is not running the valve is closed to prevent the gas from entering the intake system.

Evaporative Emission (EVAP) Canister System

The evaporative emission (EVAP) canister system prevents highly volatile fuel vapors in the fuel tank system from entering the atmosphere. The center piece of the EVAP canister system is the EVAP canister, which is filled with activated charcoal.

Activated charcoal has the ability to absorb fuel vapors, to store them and to regenerate itself when the vapors are passed on. The EVAP canister is therefore a maintenance-free component, as long as the controlling EVAP purge valve (vacuum and frequency controlled) and the connecting lines are functioning properly.

At very low temperature it is important to check whether the vent opening at the bottom of the canister is free of ice and dirt.

The EVAP canister and the fuel tank are connected by a hose in the engine compartment. When the engine is not running, the vapors from the fuel tank are fed to the EVAP canister and are stored there. When the engine is started, a second hose leads the vapors to the intake manifold to be burned in the engine. See Fig. 2-14.

METHODS TO REDUCE EMISSIONS

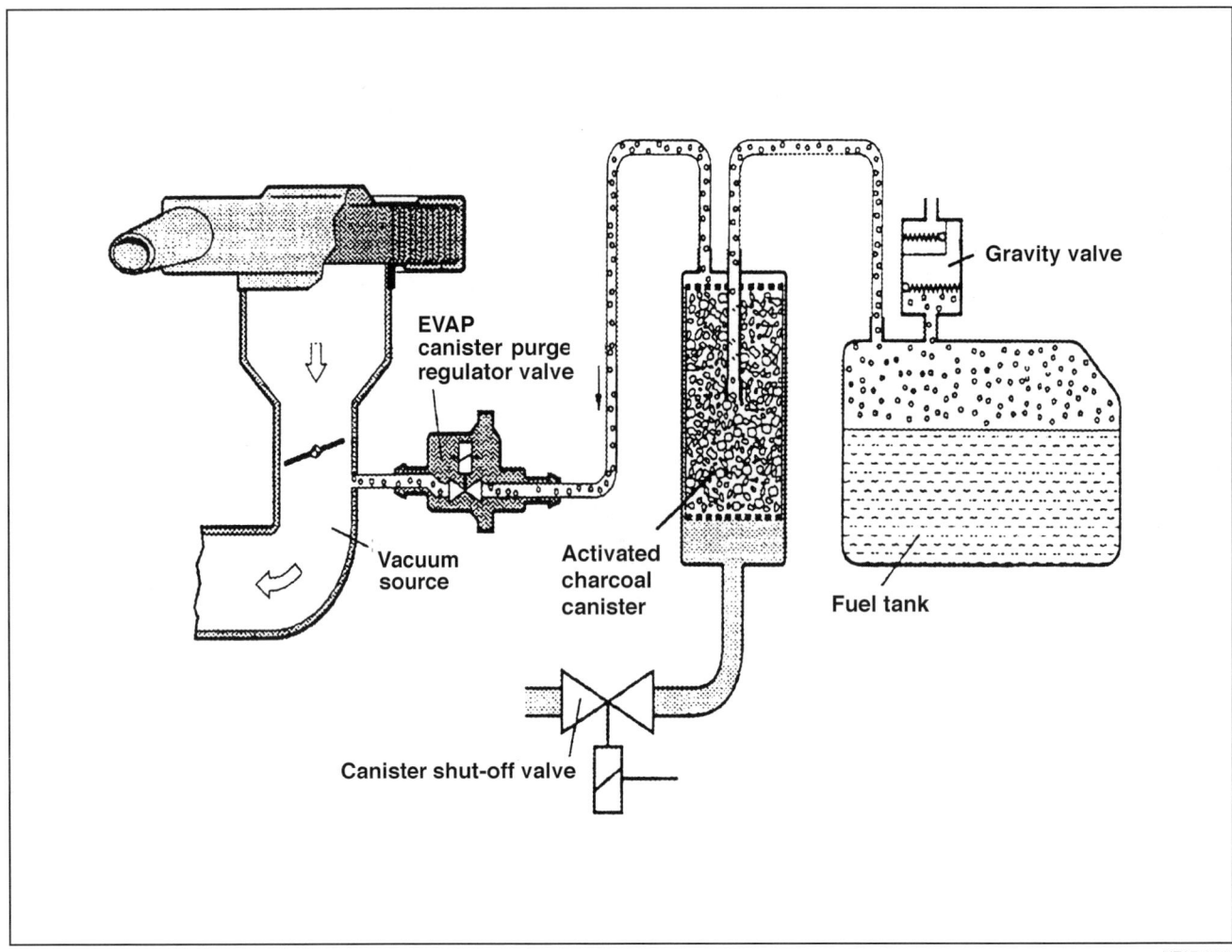

Fig. 2-14. Evaporative emission (EVAP) canister system.

Chapter 3

Engine Management Systems

General 29	**Continuous Injection System–Electronic Motronic (CIS-E Motronic)** 36
Air Flow Control (AFC) 30	Injection Part 36
Basic Factors 30	Ignition Part 36
	Basic Factors 36
Continuous Injection System (CIS)– without Lambda Regulation 32	**Digijet** 38
Basic Factors 32	Basic Factors 38
Continuous Injection System (CIS)– with Lambda Regulation 33	**Digifant**
Basic Factors 33	Injection Part 39
	Ignition Part 39
Continuous Injection System– Electronic (CIS-E) 34	Basic Factors 39
Basic Factors 34	**Motronic**
	Ignition Part 40
	Fuel Injection Part 40
	Basic Factors 40

GENERAL

This chapter is designed to provide a fundamental overview of the engine management systems used in Volkswagens during model years 1980 to 1997. This section should be used as a quick orientation reference, and not for training.

Where possible, the systems are presented in a time sequence, beginning with the oldest system first. In some instances, schematic system views are also included to show the difference between the various systems.

The system descriptions are very general because over time many components have changed, especially those concerning the emission control. We did not cover these details, however you can find the information in the applicable Volkswagen repair manual.

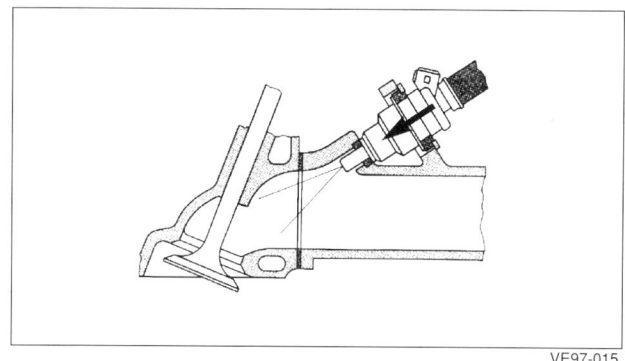

Fig. 3-1. Fuel system applications vary depending on model and model year. For a complete listing of Volkswagen models and the applicable engine management system, see **Appendix A**.

WARNING —
Never run the engine unless the work area is well ventilated. Exhaust gas emissions are extremely toxic and can kill.

Air Flow Control (AFC)

The AFC system is an electronically controlled fuel injection system. The Engine Control Module (ECM) controls the lambda regulation necessary for engine operation with little harmful emissions. See Fig. 3-2.

Engine idle can be adjusted with the idle adjustment screw. CO content can be adjusted with the CO adjustment screw in the air flow sensor. The idle air control valve provides the correct engine idle under various engine operation conditions.

Each cylinder is equipped with a solenoid-operated fuel injector. Injector opening times (injector pulse-width) are controlled by the ECM. The fuel is injected intermittently directly into the intake manifold onto the intake valves.

Basic Factors

The basic factors for the control of the fuel injection are the engine speed and the volume of the intake air that is controlled by the Mass Air Flow Sensor. These factors are used by the Engine Control Module (ECM) which calculates the opening times (pulse-width) of the fuel injectors. To optimize the fuel/air mixture regarding emissions, the ECM monitors the following factors:

- Signal of the intake air temperature sensor
- Signal of the engine coolant temperature sensor
- Signal of the throttle position switch
- Signal of the oxygen sensor

Depending on these values, the ECM influences the fuel/air mixture through the opening times of the fuel injectors.

Engine Mangement Systems

L-Jetronic

3. Engine Control Module (ECM)
4. Fuel Injector
5. Volume Air Flow Sensor
6. Engine Coolant Temperature Sensor
7. Thermal Time Switch
8. Cold Start Injector
9. Fuel pump
10. Fuel filter
11. Fuel Pressure Regulator
12. Auxiliary Air Regulator Valve
13. Throttle Position Switch
14. Combination Relay

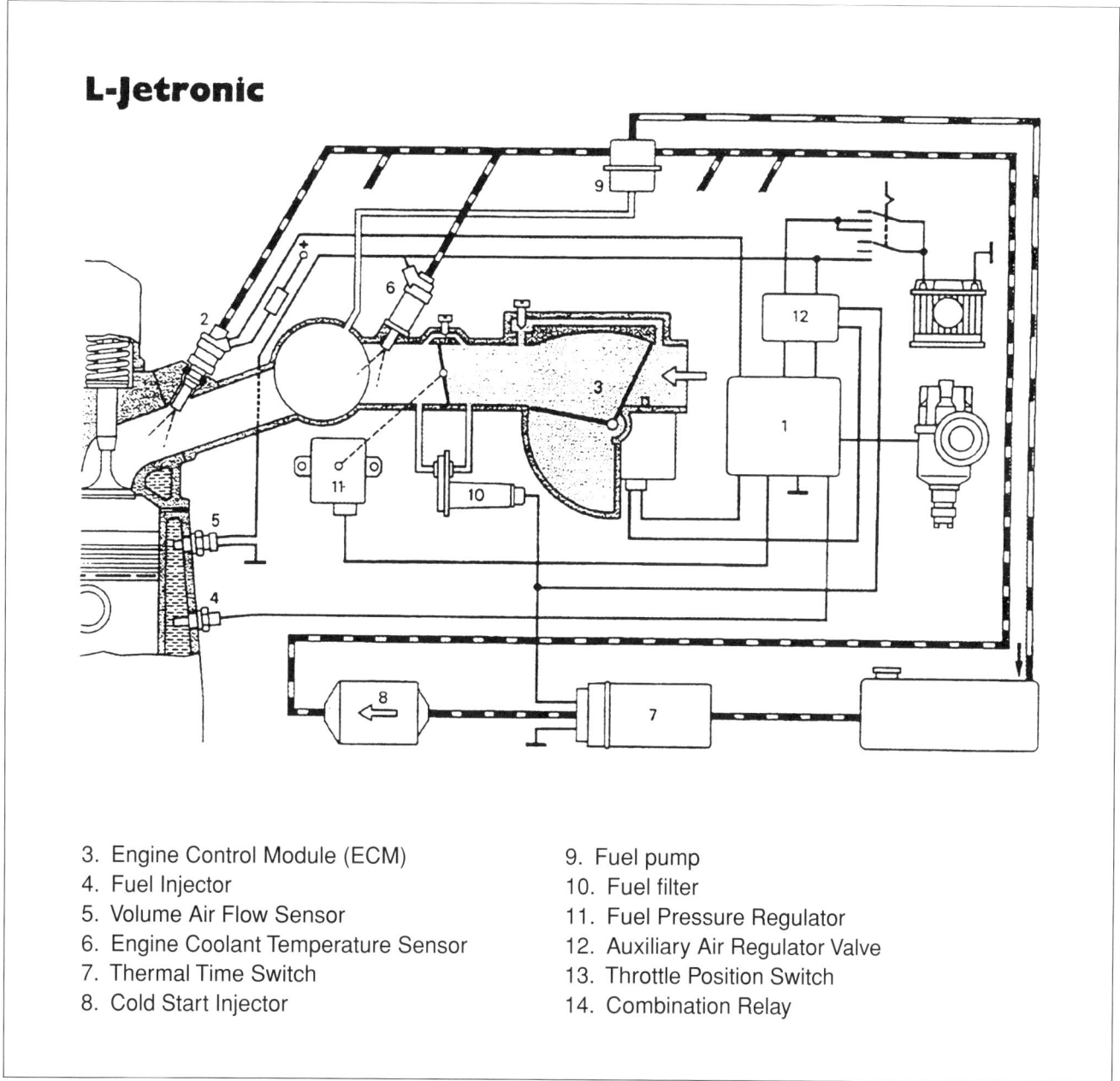

Fig. 3-2. AFC (L-Jetronic) fuel injection system.

AIR FLOW CONTROL (AFC)

Continuous Injection System (CIS) (Without Lambda Regulation)

CIS is a mechanical-hydraulic fuel system.

A sensor plate—deflected by intake air—controls the basic fuel metering function via the control plunger. The position of the control plunger controls basic fuel flow to the injectors.

Each cylinder is equipped with a mechanical fuel injector that continuously injects fuel into the intake manifold directly onto the intake valves. The injected volume of fuel depends on the position of the control plunger and on the fuel pressure.

Basic Factors

The basic factor for controlling the amount of fuel injected is the intake air.

If the engine is cold, additional fuel is injected into the intake manifold during starting by the cold start injector to assure dependable cold starts and cold idle. The cold start injector is controlled by time and temperature to inject a limited amount of fuel after a cold start. When the warm engine is started, the injector does not open.

To achieve mixture enrichment during the warm-up phase, the warm-up regulator decreases the control pressure counterforce on the top of the control plunger. This allows the air flow sensor plate to deflect further for a given volume of intake air.

The auxiliary air regulator, controlled by a bimetallic or an expansion medium supplies the engine with additional by-pass air (by circumventing the throttle). The additional air causes the air flow sensor plate to further deflect which results in an increase of the idle speed during the warm-up phase. This system does not have lambda regulation.

Engine Management Systems

CONTINUOUS INJECTION SYSTEM (CIS) (WITH LAMBDA REGULATION)

CIS is a mechanical-hydraulic fuel system with electronic controls.

A sensor plate—deflected by intake air—controls the basic fuel metering function via the control plunger. The position of the control plunger controls basic fuel flow to the injectors.

In addition, the CIS with Lambda incorporates a control unit that adapts the injected fuel via a frequency valve in the fuel return line.

Each cylinder is equipped with a mechanical fuel injector that continuously injects fuel into the intake manifold directly onto the intake valves. The injected volume of fuel depends on the position of the control plunger and on the fuel pressure.

Basic Factors

The basic factor for controlling the amount of fuel injected is the intake air. To optimize the air/fuel mixture regarding emissions, the signal of the oxygen sensor is evaluated by the Engine Control Module (ECM) to adjust the mixture accordingly.

If the engine is cold, additional fuel is injected into the intake manifold during starting by the cold start injector to assure dependable cold starts and cold idle. The cold start injector is controlled by time and temperature to inject a limited amount of fuel after a cold start. When the warm engine is started, the injector does not open.

To achieve mixture enrichment during the warm-up phase, the warm-up regulator decreases the control pressure counterforce on the top of the control plunger. This allows the air flow sensor plate to deflect further for a given volume of intake air.

The auxiliary air regulator, controlled by a bimetallic or an expansion medium supplies the engine with additional by-pass air (by circumventing the throttle). The additional air causes the air flow sensor plate to further deflect which results in an increase of the idle speed during the warm-up phase.

Continuous Injection System—Electronic (CIS–E)

The CIS-E is a mechanical-hydraulic fuel system with electronic controls.

A sensor plate—deflected by intake air—controls the basic fuel metering function via the control plunger. The position of the control plunger controls basic fuel flow to the injectors. See Fig. 3-3.

CIS-E incorporates an Engine Control Module (ECM) that adapts the injected fuel via a pressure actuator at the fuel distributor.

Each cylinder is equipped with a mechanical fuel injector that continuously injects fuel into the intake manifold directly onto the intake valves. The injected volume of fuel depends on the position of the control plunger and on the fuel pressure.

Basic Factors

If the engine is cold, additional fuel is injected into the intake manifold during starting by the cold start injector to assure dependable cold starts and cold idle. The cold start injector is controlled by time and temperature to inject a limited amount of fuel after a cold start. When the engine is warm, the injector does not open during starting.

To achieve mixture enrichment during the warm-up phase, current to the pressure actuator is increased and this increases fuel flow to the injectors.

The basic factor for the control of the fuel to be injected is the volume of the intake air. To optimize the air/fuel mixture regarding emissions and fuel consumption, the ECM evaluates the following signals or values:

- Engine speed
- Signal of the engine coolant temperature sensor
- Signal of the oxygen sensor
- Signal of the throttle position switch
- Signal of the air sensor plate potentiometer

Depending on these factors the ECM controls the mixture via the pressure actuator.

Engine Mangement Systems

1. Fuel tank
2. Fuel pump
3. Accumulator
4. Fuel filter
5. Pressure regulator
6. Air sensor plate
6a. Potentiometer
7. Fuel distributor
8. Fuel injector
9. Cold start injector
10. Thermal time switch
11. Throttle valve
12. Throttle position switch
13. Auxiliary air regulator
14. Engine coolant temperature sensor
15. Engine control module
16. Pressure actuator
17. Oxygen sensor
18. Ignition distributor
19. Control relay
20. Ignition starter switch

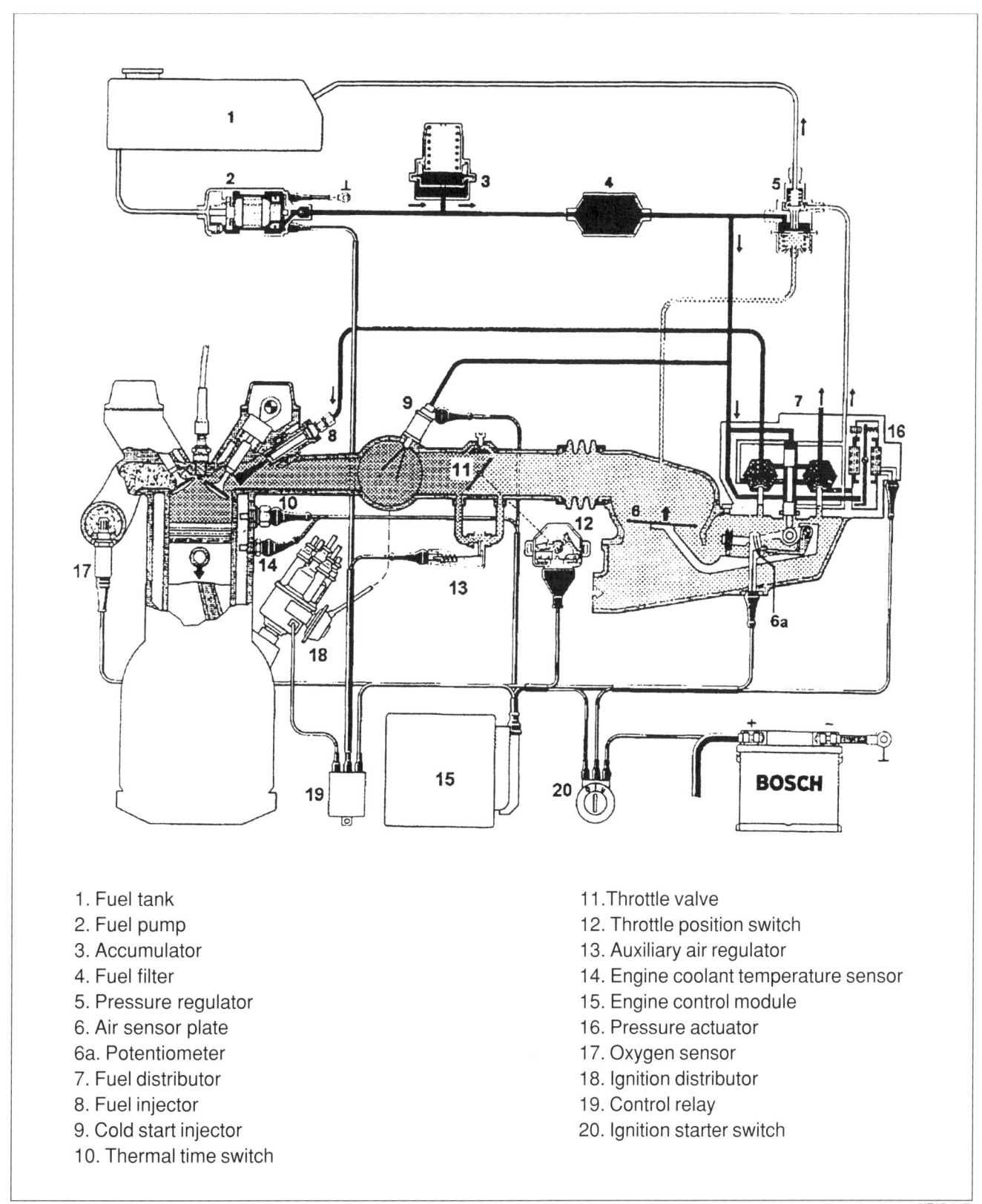

Fig. 3-3. CIS-E (KE-Jetronic) fuel injection system.

CONTINUOUS INJECTION SYSTEM—ELECTRONIC (CIS–E)

CIS-E Motronic

The CIS-E Motronic is a combined fuel injection and ignition system. The electronic Engine Control Module (ECM) combines the ignition and fuel injection functions as well as knock and Lambda regulation.

The CIS-E Motronic is capable of recognizing faults in the fuel injection and ignition systems and compensate for them in a limited way. If there is a substantial fault that cannot be compensated for, a signal will appear in the fault memory.

The adjustment of the engine idle speed is done by the ECM and the idle air control valve and is not adjustable. CO content is adjusted with the CO adjustment screw at the mixture control unit.

NOTE
Before adjusting CO, read the fault memory first to determine whether the Lambda regulation limits were exceeded.

Injection Part

Each cylinder is equipped with a mechanical fuel injector that continuously injects the fuel into the intake manifold directly onto the intake valves.

A sensor plate—deflected by intake air—controls the basic fuel metering function via the control plunger. The position of the control plunger controls basic fuel flow to the injectors.

The ECM controls the injection volume via the pressure actuator at the fuel distributor. When the engine is cold, additional fuel is injected into the intake manifold by the cold start injector to assure dependable cold engine starts and cold idle.

Ignition Part

The ignition is map controlled, meaning that the criteria of fuel consumption, emission, and performance are optimally coordinated and electronically stored in memory in the form of tables (maps). The ignition map was determined on an engine test stand and provides the best ignition timing for the various engine load conditions. Ignition timing is not adjustable.

Basic Factors

The basic factor for the control of fuel injection functions is the volume of the intake air. To produce the best possible air/fuel mixture and ignition spark control, the ECM evaluates the following signals or values:

- Engine speed
- Signal of the engine coolant temperature sensor
- Signal of the air sensor plate potentiometer
- Signal of the knock sensors
- Signal of the idle air control switch
- Signal of the oxygen sensor

Depending on these factors, the ECM controls the mixture via the pressure actuator. Ignition timing (dwell angle) is also controlled via the ECM.

Engine Mangement Systems

DIGIJET

Digijet is an electronically controlled fuel injection system with an injection map. The electronic Engine Control Module (ECM) controls the lambda regulation necessary for the operation of the engine with low emissions.

Engine idle can be adjusted with the idle adjustment screw and CO content can be adjusted with the CO adjustment screw. The Idle Air Control Valve provides for the correct idle speed under various engine operating conditions.

Each cylinder is equipped with a solenoid-operated fuel injector with the opening time (pulse-width) controlled by the ECM. The fuel is intermittently injected into the intake manifold directly onto the intake valves.

Basic Factors

The basic factors for the control of the fuel injection system are engine speed and the volume of the intake air that is sensed by a Mass Air Flow Sensor. With these factors, the ECM calculates the opening time of the fuel injectors from the injection map.

To optimize the air/fuel mixture regarding emissions and fuel consumption, the Engine Control Module evaluates the following signals:

- Signal of the intake air temperature sensor
- Signal of the engine coolant temperature sensor
- Signal of the throttle position sensor
- Signal of the oxygen sensor

Depending on these factors the ECM controls the mixture via the opening time of the fuel injectors.

DIGIFANT

Digifant is a combined fuel injection and ignition system. The electronic Engine Control Module (ECM) combines the ignition and fuel injection functions as well as idle air control, knock control, and Lambda regulation. See Fig. 3-4.

NOTE
- *Older Digifant systems allowed for an adjustment of CO content and idle speed.*
- *From Jetta, MY1991 for California, a throttle position sensor was installed. On this system the adjustments are done by the ECM and only checking of the values is possible.*

Injection Part

Digifant is an electronically controlled fuel injection system with programmed injection map, idle air control and deceleration fuel shut-off.

Each cylinder is equipped with a solenoid-operated fuel injector with the opening time controlled by the ECM. The fuel is intermittently injected into the intake manifold directly onto the intake valves. The load condition is measured by the Mass Air Flow Sensor.

Idle stabilization is accomplished by a separate Idle Air Control Valve that is controlled by the ECM.

Ignition Part

The ignition is map controlled, meaning that the criteria of fuel consumption, emission, and performance are coordinated in the best possible way and electronically stored in memory in the form of tables (maps). The ignition map was determined on an engine test stand and provides the best ignition timing for the various engine load conditions.

The vehicle's ECM also controls the idle speed via the ignition timing and causes the integrated idle stabilization function to increase idle when additional electrical consumers are switched on.

Knock sensors also help to obtain optimal fuel efficiency.

Basic Factors

Assisted by the maps, the ECM needs information about the basic factors of engine speed and engine load to control fuel injection and ignition.

The data of the maps are enhanced by the following data:

- Signal of the intake air temperature sensor
- Signal of the engine coolant temperature sensor
- Signal of the throttle position sensor
- Signal of the oxygen sensor

Engine Mangement Systems 39

1. Fuel tank
2. Fuel pump
3. Ignition coil
4. Mass air flow sensor
5. Intake air temperature sensor
6. Throttle position sensor
7. Idle air control valve
8. Knock sensor
9. Engine coolant temperature sensor
10. Ignition distributor
11. Oxygen sensor
12. Fuel injector

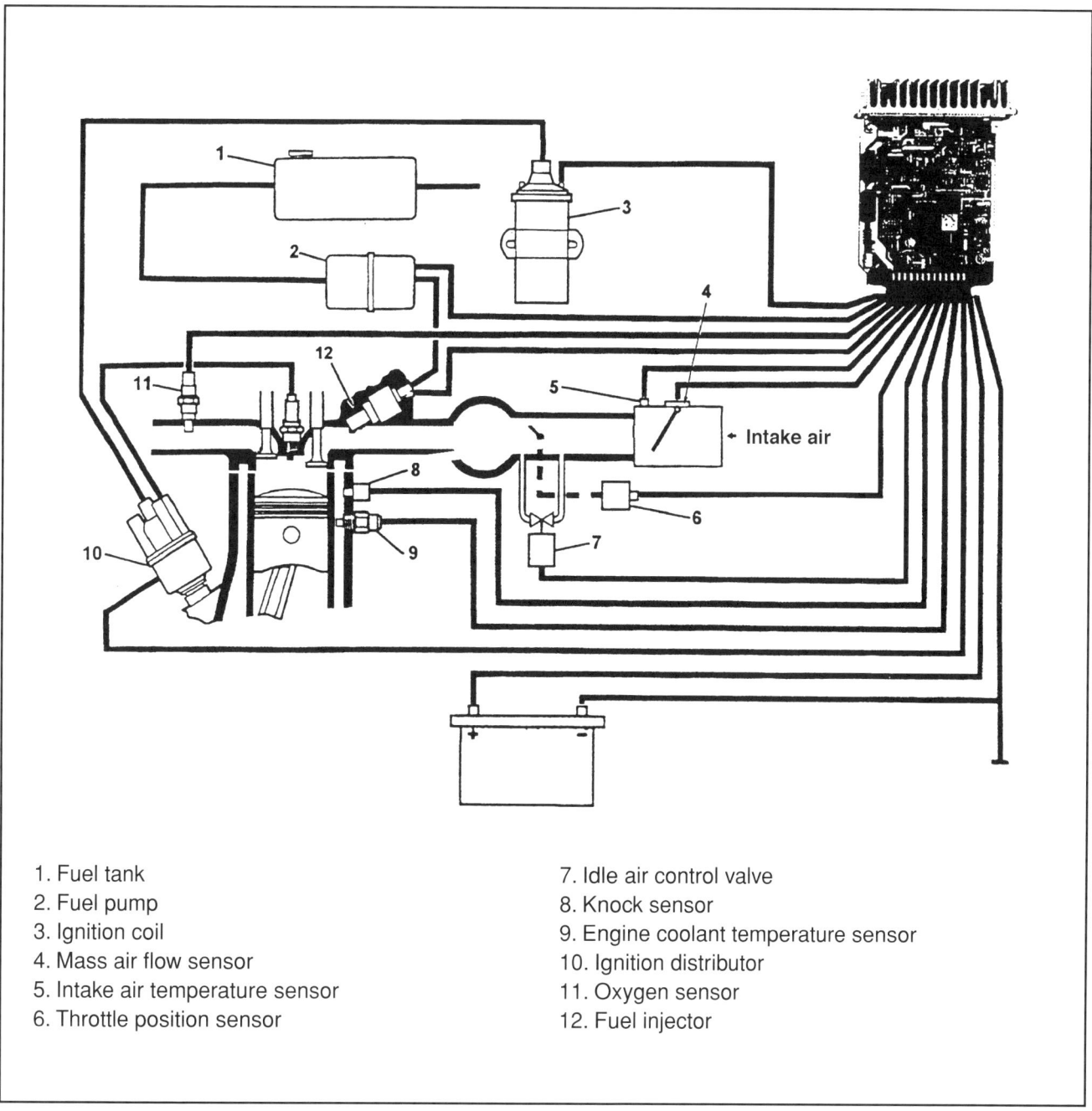

Fig. 3-4. Digifant engine management system.

DIGIFANT

MOTRONIC

Motronic is a combined fuel injection and ignition system. The electronic Engine Control Module (ECM) combines the ignition and fuel injection functions as well as idle air control, knock control and Lambda regulation. See Fig. 3-5.

Motronic is capable of recognizing faults in the fuel injection and ignition systems and compensate for them in a limited way. If there is a substantial fault that cannot be compensated for, a signal will appear in the fault memory.

Contrary to other engine management systems, CO content, engine idle speed and ignition timing cannot be adjusted.

Ignition Part

The ignition is map-controlled to maintain the best possible ignition timing adjustment under all operating conditions.

Map-controlled means that the criteria of fuel consumption, emission, and performance are best coordinated and electronically stored in memory in form of tables (maps). The ignition map was determined on an engine test stand and provides the best ignition timing for the various engine load conditions. The ECM also controls the idle speed via the ignition timing.

Fuel Injection Part

The fuel injection system is also map-programmed. Each cylinder is equipped with a solenoid-operated fuel injector with the opening time controlled by the ECM. The fuel is intermittently injected into the intake manifold directly onto the intake valves.

Basic Factors

The basic factors for the control of ignition and fuel injection are the load condition and the engine speed that is established by the Mass Air Flow Sensor. With these factors the ECM determines the ignition timing and the opening times of the fuel injectors. The data of the ECM is further enhanced by the following values:

- Signal from the intake air temperature sensor
- Signal of the engine coolant temperature sensor
- Signal of the oxygen sensor
- Signal from the knock sensor

Engine Mangement Systems 41

1. EVAP canister
2. Shut-off valve
3. Purge valve
4. Fuel pressure regulator
5. Fuel injector
6. Pressure regulator
7. Ignition coil
8. Stage sensor
9. Secondary air pump
10. Secondary air valve
11. Mass air flow sensor
12. Engine control module
13. Throttle position sensor
14. Closed throttle position switch
15. Intake air temperature sensor
16. EGR valve
17. Fuel filter
18. Knock sensor
19. Engine speed sensor
20. Engine coolant temperature sensor
21. Oxygen Sensor
22. Diagnostic interface
23. Malfunction indicator lamp
24. Differential pressure sensor
25. Fuel pump

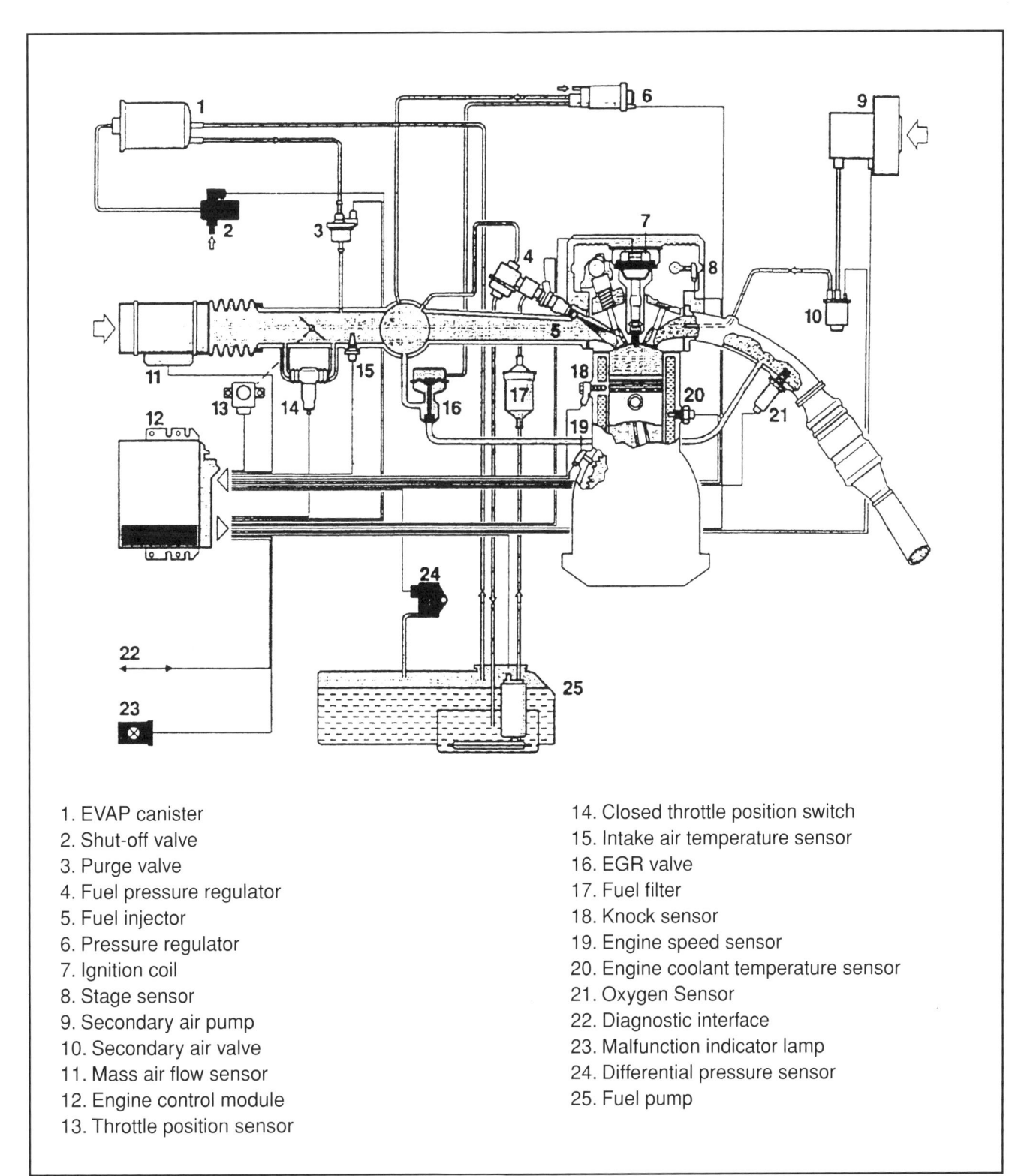

Fig. 3-5. Motronic engine management system.

MOTRONIC

Chapter 4

Troubleshooting Chart

General 43	Troubleshooting............................46
Working with the	**Pressure Test**............................47
Troubleshooting Chart 43	Method A....................................47
Troubleshooting Chart 44	Method B....................................48
Evaporative Emissions	**Checking Dual Bed Catalytic**
Control System......................... 46	**Converters**...............................49
Visual check............................ 46	Table to check function of catalytic converters 49

GENERAL

This chapter contains the Inspection and Maintenance (I/M) Troubleshooting Chart. It is recommended that the troubleshooting program begin with the Troubleshooting Chart found here.

Also included in this chapter is diagnostic and testing information for the Evaporative Emissions Control (EVAP) System and the Catalytic Converter.

NOTE

The troubleshooting chart refers to the Volkswagen fault read out unit VAG 1551. It is helpful if the servicing technician has access to this tool when servicing late model (most 1990 and later) cars. The tool is used to access stored Diagnostic Trouble Codes (DTCs) in the On-board Diagnostic (OBD) System.

WORKING WITH THE TROUBLESHOOTING CHART

The troubleshooting chart consists of three main branches from which the chart can be entered:

1. Evaporative Control System
2. Emissions
3. Safety Inspection

These branches conform to the layout an I/M test report to a certain degree. (Remember that a typical I/M test consists of a visual inspection, a tailpipe emissions test, an evaporative emission control system purge, and fuel tank pressure test.)

NOTE

*Before starting to check or repair a vehicle that has failed an I/M test, refer to the **Customer Check List** found in Chapter 1. This list may already contain important information as to why the vehicle failed the test.*

The I/M test report provided by a Test Station will most likely list the reason(s) for failing the test. The main topic on the report usually identifies the reason for the failure. This topic should match one of the main branches on the troubleshooting chart.

For Example: According to the test report, the vehicle has failed "Emissions*" (it is assumed here that the vehicle has passed all other tests).

Enter the troubleshooting chart at "Emissions" and follow each step. Disregard the other branches for this example. The troubleshooting chart refers to parts of the Volkswagen service literature where detailed repair and maintenance instructions can be found.

*Use of these terms is only for orientation. The actual terms may be different on the test report.

TROUBLESHOOTING CHART

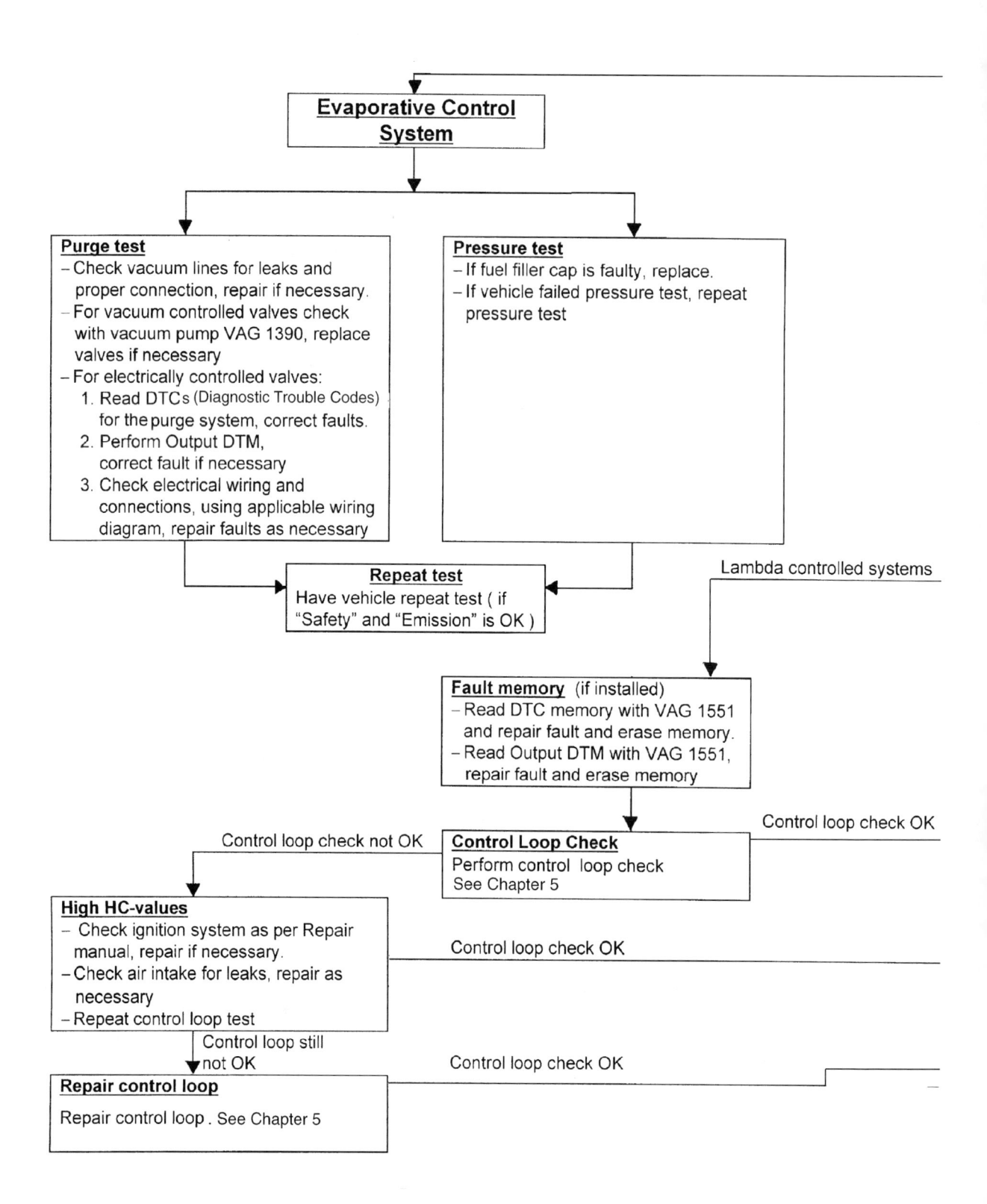

Evaporative Control System

Purge test
- Check vacuum lines for leaks and proper connection, repair if necessary.
- For vacuum controlled valves check with vacuum pump VAG 1390, replace valves if necessary
- For electrically controlled valves:
  1. Read DTCs (Diagnostic Trouble Codes) for the purge system, correct faults.
  2. Perform Output DTM, correct fault if necessary
  3. Check electrical wiring and connections, using applicable wiring diagram, repair faults as necessary

Pressure test
- If fuel filler cap is faulty, replace.
- If vehicle failed pressure test, repeat pressure test

Repeat test
Have vehicle repeat test (if "Safety" and "Emission" is OK)

Lambda controlled systems

Fault memory (if installed)
- Read DTC memory with VAG 1551 and repair fault and erase memory.
- Read Output DTM with VAG 1551, repair fault and erase memory

Control Loop Check
Perform control loop check
See Chapter 5

Control loop check not OK · Control loop check OK

High HC-values
- Check ignition system as per Repair manual, repair if necessary.
- Check air intake for leaks, repair as necessary
- Repeat control loop test

Control loop still not OK

Repair control loop
Repair control loop. See Chapter 5

Control loop check OK

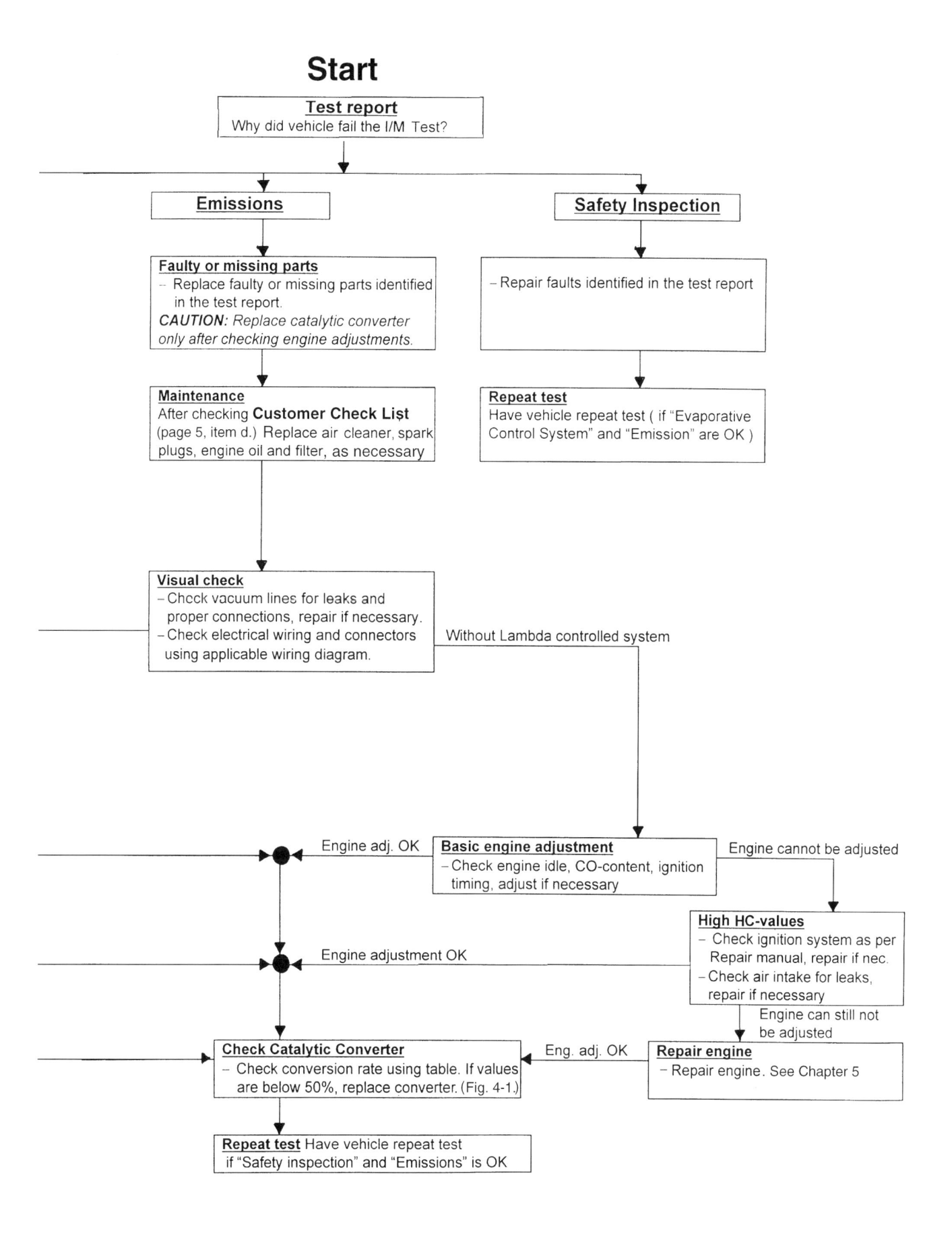

Troubleshooting Chart

Start

Test report
Why did vehicle fail the I/M Test?

↓ ↓

Emissions

Faulty or missing parts
– Replace faulty or missing parts identified in the test report.
CAUTION: Replace catalytic converter only after checking engine adjustments.

↓

Maintenance
After checking **Customer Check List** (page 5, item d.) Replace air cleaner, spark plugs, engine oil and filter, as necessary

↓

Visual check
– Check vacuum lines for leaks and proper connections, repair if necessary.
– Check electrical wiring and connectors using applicable wiring diagram.

Without Lambda controlled system →

Safety Inspection

– Repair faults identified in the test report

↓

Repeat test
Have vehicle repeat test (if "Evaporative Control System" and "Emission" are OK)

Engine adj. OK ← **Basic engine adjustment**
– Check engine idle, CO-content, ignition timing, adjust if necessary
→ Engine cannot be adjusted

↓

High HC-values
– Check ignition system as per Repair manual, repair if nec.
– Check air intake for leaks, repair if necessary

Engine adjustment OK ←

Engine can still not be adjusted ↓

Check Catalytic Converter
– Check conversion rate using table. If values are below 50%, replace converter. (Fig. 4-1.)
← Eng. adj. OK — **Repair engine**
– Repair engine. See Chapter 5

↓

Repeat test Have vehicle repeat test if "Safety inspection" and "Emissions" is OK

TROUBLESHOOTING CHART

EVAPORATIVE EMISSIONS CONTROL SYSTEM

A part of the I/M Emission Test is the "EVAP Test" to check the fuel tank system for leaks. The hoses on older vehicles deteriorated by ozone-cracks and aging are often the reason why these vehicles do not pass the EVAP Test.

The EVAP Test consists of three parts:

1. The pressure test to check for leaks in the fuel tank system
2. The purge test to check the function of the EVAP canister system
3. Gas cap test

If a vehicle has failed the EVAP Test, first check the test report of the emissions Test Station whether the vehicle has also failed the purge test, the pressure test or the gas cap test.

If the vehicle failed the gas cap test, replace the gas cap and return the vehicle for a repeat test if no other faults were listed on the test report.

If other faults are listed, refer to the troubleshooting chart for additional troubleshooting information.

Visual check

If the vehicle failed the purge or pressure test, visually check the vacuum lines of the fuel tank system for correct connection and whether the hoses are in sound condition.

If hoses are disconnected or not properly connected make the necessary repairs. Afterwards perform a pressure test (see **Pressure Test**, given later). If the pressure test is now OK, have the vehicle return to the Test Station for a repeat test.

Troubleshooting

Newer vehicles as of model year 1990 (engine codes AAA, ABA, ACU, AAF and 9A) are equipped with On Board Diagnostic (OBD) systems, as part of the engine management system. Be sure to "read" the Diagnostic Trouble Code fault memory for EVAP system faults and repair as necessary.

On later cars, check the electrical functioning of the EVAP canister valve. See Volkswagen Repair Manual: *On-board Diagnostics Troubleshooting, Repair Group 01*.

For electrically operated purge regulator valves, connect the hose from the EVAP canister to the valve. When you hear the cycling of the valve, blow into the hose. The valve should be closed during repeated blowing. If the valve remains open, replace the valve. If no other faults were listed, have the vehicle return to the test station for a repeat test. Replace the valve if necessary. See Volkswagen Repair Manual: *General, Engine, Repair Group 26*.

On early cars with a vacuum-operated valve, check the valve using a hand vacuum pump. See Volkswagen Repair Manual: *General, Engine, Repair Group 26*. Replace the valve if necessary.

PRESSURE TEST

If the EVAP canister valves function properly, perform a pressure test on the entire fuel tank system, especially the vacuum hoses. Use an EVAP Emission Tester such as US 4487 or EEEA 102A Evaporative Emission Tester by SUN and a four-gas analyzer such as VAG 1788 to make the test.

Before starting the pressure test be sure that the tank is at least 2/3 full, otherwise it will take a long time (and a lot of nitrogen) to pressurize the system. For more information about the use of the testers see the manufacturer's instructions.

If the vehicle passes the pressure test and there are no other faults listed, have the vehicle return to the emission Testing Station for a repeat test.

If the fuel tank system cannot be pressurized, there are two methods for finding leaks.

Method A

1. Apply soapy water to all hoses and connections.

2. Check the pressure in the tank system and pressurize again.
   - Parts of the pressurized system will show small bubbles. Hoses showing bubbles should be replaced. If you see bubbles on connectors, loosen the connectors and tighten them again.

3. If necessary, release the test pressure as described in the applicable Volkswagen Repair Manual and repeat the pressure test.
   - If the pressure holds and no other faults were listed in the report, have the vehicle return to the Testing Station for a repeat test.

Method B

1. With engine at operating temperature (engine oil temperature at least 80 °C / 176 °F), allow it to run at idle.

   > **WARNING—**
   > Never run the engine unless the work area is well ventilated. Exhaust gas emissions are extremely toxic and can kill.

2. Properly lift the vehicle on a lift.

   > **WARNING—**
   > Never work under a lifted car unless it is solidly supported on stands or lift equipment designed for the purpose. Do not support a car on cinder blocks, hollow tires or other props that may crumble under continuous load. Never work under a car that is supported solely by a jack.

3. To detect leaks, use the probe of a four-gas analyzer to check the entire fuel system. Check the following areas:
   - Entire air intake system from the throttle housing to the fuel tank, including the gas cap, and the EVAP canister.

4. If high HC values are displayed in one particular area, a leak is close by. Hoses that show high HC readings have become porous and must be replaced. If you find leaking connectors, disconnect and reconnect, then check again. If the gas cap is leaking, it should be replaced.

   > **NOTE**
   > If high values are read in the area of the rear muffler, they may be false readings if the exhaust gas is not properly vented and leaking into the area. Correct the venting problem first before proceeding with the leak test.

5. After repairs, repeat the pressure test. If the pressure holds and no other faults are listed in the test report, have the vehicle return to the emission Test Station for a repeat test.

CHECKING DUAL-BED CATALYTIC CONVERTERS

Use the table on the following page when checking the function of the catalytic converter.

1. Connect tester VAG 1363, or an equivalent a four-gas analyzer such as VAG 1788 to the CO test connection on the exhaust (pre-catalytic converter).

   NOTE
   The CO test connection (test tap) is part of the exhaust system and is positioned before the catalytic converter.

2. Run engine at approximately 2500 rpm for at least one minute.

3. With the engine at a constant rpm, measure the pre-catalytic converter CO-content at CO test connection. Make note of the result (A-value in the table). See Fig. 4-1.

4. Measure the post-catalytic converter CO-content at tail pipe (engine should be held at same speed as in step 3 above). Make note of the result (B-value in the table). See Fig. 4-1.

5. Read off percentage value from the table. See Fig. 4-1.
   - If value is 50% or higher (white area in table), the converter is OK and does not need to be replaced.
   - If value is below 50% (dark area in table), the converter is faulty and should be replaced.

CHECKING DUAL-BED CATALYTIC CONVERTERS

50 Chapter 4

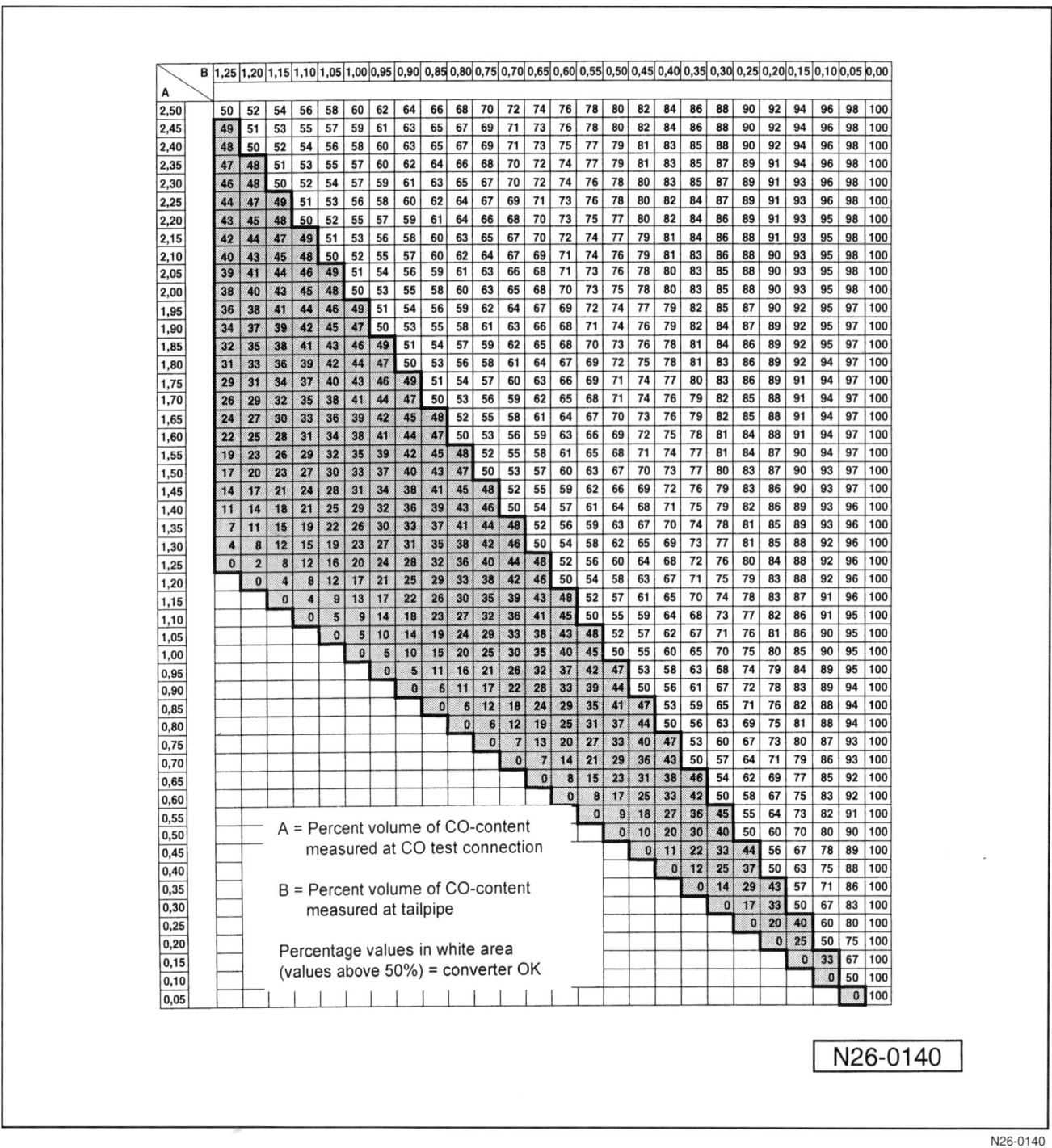

Fig. 4-1. Table to check function of catalytic converters (only for vehicles with regulated converters).

A = Percent volume of CO-content measured at CO test connection

B = Percent volume of CO-content measured at tailpipe

Percentage values in white area (values above 50%) = converter OK

CHECKING DUAL-BED CATALYTIC CONVERTERS

Chapter 5

Emission Diagnosis and Repair

GENERAL 51	Engine Code GX 81
Finding Volkswagen Engine Codes 51	Engine Code HT 84
Engine Code AAA 52	Engine Code JH 87
Engine Code AAF 55	Engine Code JN 90
Engine Code ABA 58	Engine Code KX 93
Engine Code ABG 62	Engine Code MV 96
Engine Code 9A 65	Engine Code PF 99
Engine Code 2H 68	Engine Code PG 102
Engine Code CV 71	Engine Code PL 106
Engine Code DH 74	Engine Code RD 109
Engine Code EJ 77	Engine Code RV 112
Engine Code EN 78	Engine Code UM 90

GENERAL

This chapter is arranged alphabetically by engine code.

NOTE
- *A listing of Volkswagen models and applicable engine codes can be found in **Appendix A**.*
- *It is recommended that the technician have access to the Volkswagen fault read out unit (VAG 1551) when servicing late model (most 1990 and later) cars.*
- *Some Volkswagen models may have been available with more than one engine code during a given model year. Therefore, it is recommended that the engine code be verified by locating the code on the engine itself.*

Finding Volkswagen Engine Codes

The engine code that identifies an engine can be viewed in the engine compartment. The code is usually stamped on a flat area on the top of the engine block. See Fig. 5-1.

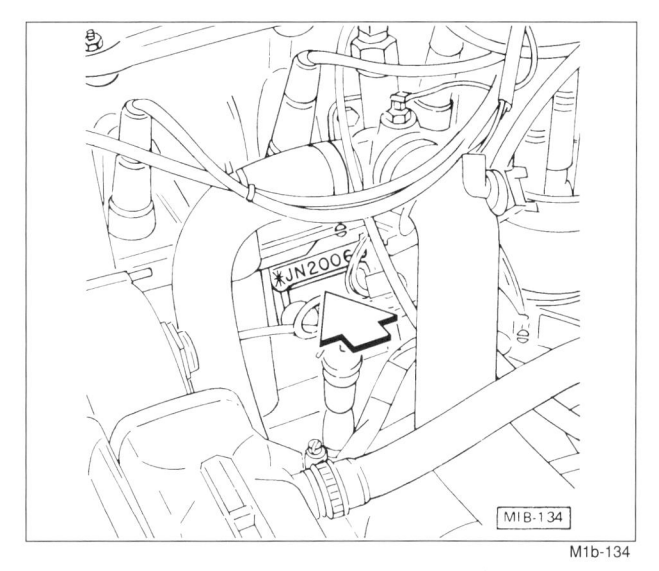

Fig. 5-1. Typical location of VW engine code (arrow).

The engine code may also be located in other places on the vehicle. On late models, the code can be found on a sticker on the timing belt cover or on the vehicle identification stamping in the luggage compartment.

52 Chapter 5

ENGINE CODE AAA
(2.8 liter/6 cylinders, Motronic)

Engines of vehicles listed below have the engine code **AAA** and **Motronic** engine management system.

Engine displacement: 2.8 liter/6 cylinders. The engine is equipped with the emission influencing components listed in **Table a**.

Model	Model Year
Corrado / SLC	1992, 1993, 1994
Passat GLX	1993, 1994, 1995, 1996, 1997
Golf GTI VR6	1994, 1995, 1996, 1997
Jetta GLX	1994, 1995, 1996, 1997

Table a. Emission Influencing Components: Engine Code AAA

Fuel System	Motronic
Ignition System	Ignition distributor / without ignition distributor[1], ECM
Sensors	Mass Air Flow Sensor
	Intake Air Temperature Sensor
	Engine Coolant Temperature Sensor (NTC2)
	Heated Oxygen Sensor[2]
	Throttle Position Sensor
	2 Knock Sensors
	Crankshaft Position Sensor (Hall effect)[3]
	RPM Sensor
Control Devices	EVAP canister purge via frequency valve
	EVAP canister: located in right front wheel housing
	Vacuum controlled crankcase ventilation
	Three-Way Catalytic Converter
	Idle speed control via Idle air control valve
	Exhaust gas recirculation (EGR): EGR Control Solenoid Valve, EGR Valve, EGR Temp. Sensor [5]
	Secondary Air Injection (AIR): AIR Valve, Solenoid AIR Control Valve[4]
	Diagnostic trouble code, MIL Lamp
Other Components	Inlet restrictor

[1] As of MY 1992/93, the distributor ignition was changed to non-distributor ignition
[2] Corrado, MY 1994 and MY 1996/97 models equipped with two oxygen sensors
[3] Cars with distributor ignition only
[4] Corrado SLC, Passat GLX, MY 1993 and Corrado SLC, MY 1992 without secondary air injection
[5] Passat GLX, MY 1993 without EGR

ENGINE CODE AAA (2.8 liter/6 cylinders, Motronic)

Emission Diagnosis and Repair

Lambda control loop, checking

Before checking the control loop, be sure that you used the Troubleshooting Chart up to this point. Read Diagnostic Trouble Code (DTC) memory and Output Diagnostic Test Mode (DTM) using the VAG 1551 fault read out unit. If faults are found, correct them first and then clear the fault memory. See applicable Volkswagen Repair Manual.

Test conditions

- Engine oil temperature must be at least 80°C (176°F).
- All electrical consumers such as air conditioner, radio, fan, etc. must be switched off.

1. Connect the following test equipment:
   - Volkswagen fault read out unit, VAG 1551
   - Ignition tester such as VAG 1367
   - Four-gas analyzer, such as VAG 1788 to tailpipe.

2. Run engine at operating temperature.

3. Increase engine speed to 2500 to 2800 rpm for at least 30 seconds until the Lambda value has stabilized.
   - Read first Lambda value and make note of it.

4. Reduce engine speed to idle and wait at least 30 seconds until Lambda value has stabilized again.
   - Read second Lambda value and make note of it.

5. Remove crankcase ventilation hose at the cylinder head connection. See Fig. 5-2. Close off hose and open again after 30 seconds. The Lambda value has to increase briefly and then settle again at the previous value.
   - If the Lambda value did not change, the control loop does not operate properly. In this case reconnect the hose and continue with the Troubleshooting Chart at "Control loop not OK".
   - If the Lambda value changes, continue checking as follows.

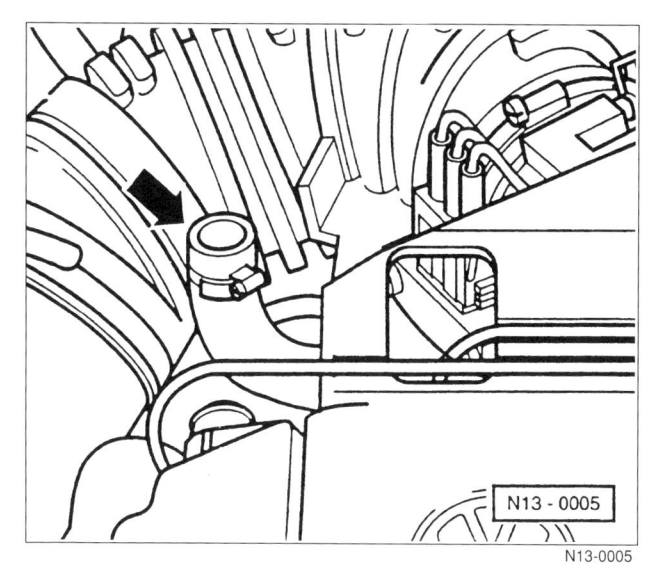

Fig. 5-2. Crankcase ventilation hose (arrow) to be disconnected from cylinder head connection.

6. After 60 seconds read the third Lambda value and make note of it.

7. Reinstall crankcase ventilation hose and secure.

8. Check if air conditioner, power steering or automatic transmission affect engine performance.

9. Evaluate noted Lambda values.
   - If one or more of the noted Lambda values are out of the range 0.97 to 1.03, or if the Lambda value does not stabilize, the control loop does not function properly. Continue in the Troubleshooting Chart at "Control loop not OK".
   - If the noted Lambda values are within 0.97 to 1.03 and if the Lambda value stabilizes, the control loop is OK. Continue in the Troubleshooting Chart at "Control loop OK".

NOTE

To see the relationship between the Lambda value and oxygen sensor output voltage, see Fig. 2-13 in Chapter 2.

ENGINE CODE AAA (2.8 liter/6 cylinders, Motronic)

Control loop, repairing

Equipment needed:

- Ignition tester such as VAG 1367
- CO tester such as VAG 1363 or SUN 105

1. Perform basic engine adjustment as per applicable Volkswagen Repair Manual.

2. If the control loop is still not OK, continue troubleshooting as follows.

Checking and Repairing Emission Influencing Components

Check the following components in the table. Perform individual checking steps as per applicable Volkswagen Repair Manual.

NOTE
- *Components that were already recognized as being faulty and were repaired do not have to be rechecked or replaced.*
- *Should you however have found a fault and repaired it, re-check the control loop once again. If the control loop checks out OK, continue in the Troubleshooting Chart with "Control loop OK".*

Additional Components to Test	Volkswagen Repair Manual (Repair manual title and applicable repair group)
Fuel pressure regulator	**Fuel Injection and Ignition Repair Group 24**
Idle Air Control Valve	
Throttle Position Switch	
Intake Air Temperature Switch	
Fuel Injectors	
EGR system	**General, Engine Repair Group 26**
Knock Sensors I and II	**Fuel Injection and Ignition Repair Group 28**
Camshaft Position Sensor	

ENGINE CODE AAA (2.8 liter/6 cylinders, Motronic)

Engine Code AAF
(2.5 liter/5 cylinders, Digifant)

Engines of vehicles listed below have the engine code **AAF** and **Digifant** engine management system.

Engine displacement: 2.5 liter/5 cylinders. The engine is equipped with the emission influencing components listed in **Table b**.

Model	Model Year
EuroVan	1992, 1993, 1994

Table b. Emission Influencing Components: Engine Code AAF

Fuel System	Digifant - MFI
Ignition System	Ignition distributor, Engine control module (ECM)
Sensors	Manifold absolute pressure sensor
	Intake Air Temperature Sensor (NTC1)
	Engine Coolant Temperature Sensor (NTC2)
	Heated Oxygen Sensor
	Throttle Position Sensor
Control Devices	EVAP canister purge via frequency valve
	Vacuum controlled crankcase ventilation
	Three-Way Catalytic Converter
	Idle speed control via Idle air control valve
Other Components	Inlet restrictor

56   Chapter 5

Lambda control loop, checking

Before checking the control loop, be sure be sure that you used the Troubleshooting Chart up to this point. Read Diagnostic Trouble Code (DTC) memory and Output Diagnostic Test Mode (DTM) using the VAG 1551 fault read out unit. If faults are found, correct them first and then clear the fault memory. See applicable Volkswagen Repair Manual.

Test conditions

- Engine oil temperature must be at least 80°C (176°F).
- All electrical consumers such as air conditioner, radio, fan, etc. must be switched off.

1. Connect the following test equipment:
   - Volkswagen fault read out unit, VAG 1551
   - Ignition tester such as VAG 1367
   - Four-gas analyzer, such as VAG 1788 to CO test connection on exhaust manifold

2. Run engine at operating temperature.

3. Increase engine speed to 2500 to 2800 rpm for at least 30 seconds until the Lambda value has stabilized.
   - Read first Lambda value and make note of it.

4. Reduce engine speed to idle and wait at least 30 seconds until Lambda value has stabilized again.

5. Read second Lambda value and make note of it.

6. Remove vacuum hose at the fuel pressure regulator. See Fig. 5-3. The Lambda value has to increase briefly and then settle again at the previous value.
   - If the Lambda value did not change, the control loop does not operate properly. In this case reconnect the vacuum hose and continue with the Troubleshooting Chart at "Control loop not OK".
   - If the Lambda value changes, continue checking as follows.

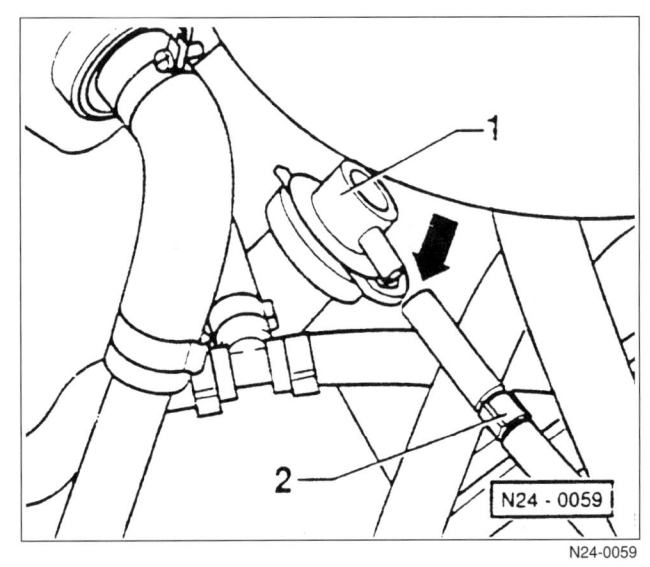

Fig. 5-3. Vacuum hose (arrow) to be disconnected from fuel pressure regulator.

7. After 60 seconds make note of the third Lambda value.

8. Reinstall vacuum hose and secure. Reconnect all hoses or lines that were disconnected.

9. Check if air conditioner, power steering or automatic transmission affect engine performance.

10. Evaluate noted Lambda values.
    - If one or more of the noted Lambda values are out of the range 0.97 to 1.03, or if the Lambda value does not stabilize, the control loop does not function properly. Continue in the Troubleshooting Chart at "Control loop not OK".
    - If the noted Lambda values are within 0.97 to 1.03 and if the Lambda value stabilizes, the control loop is OK. Continue in the Troubleshooting Chart at "Control loop OK".

NOTE

To see the relationship between the Lambda value and oxygen sensor output voltage, see Fig. 2-13 in Chapter 2.

ENGINE CODE AAF (2.5 liter/5 cylinders, Digifant)

Emission Diagnosis and Repair

Control loop, repairing

Equipment needed:

- Ignition tester such as VAG 1367
- Trigger plies such as VAG 1367/8
- CO tester such as VAG 1363

1. Perform basic engine adjustment as per applicable Volkswagen Repair Manual.
   - Check engine idle speed, adjust if necessary (not adjustable from 7/92)
   - Check CO content, adjust if necessary (not adjustable from 7/92)
   - Check ignition timing, adjust if necessary

2. If the control loop is still not OK, continue troubleshooting as follows.

Checking and Repairing Emission Influencing Components

Check the following components listed in the table. Perform individual checking steps as per applicable Volkswagen Repair Manual.

NOTE
- *Components that were already recognized as being faulty and were repaired do not have to be rechecked or replaced.*
- *Should you however have found a fault and repaired it, re-check the control loop once again. If the control loop checks out OK, continue in the Troubleshooting Chart with "Control loop OK."*

Additional Components to Test	Volkswagen Repair Manual (Repair manual title and applicable repair group)
Oxygen sensor and Lambda regulation	**Fuel Injection and Ignition Repair Group 24**
Intake Air Temperature Switch	
Idle Air Control Valve	
Cold Start Injector	
Fuel Pressure Regulator	
Fuel Injectors	
Engine Coolant Temperature Sensor	
Throttle Position Switch	
Closed Throttle Position Switch	
Acceleration Enrichment, Full Throttle Switch and Deceleration Fuel Shut-Off	
Manifold Absolute Pressure Sensor	**General, Engine Repair Group 26**
Camshaft Position Sensor	**Fuel Injection and Ignition Repair Group 28**

ENGINE CODE AAF (2.5 liter/5 cylinders, Digifant)

Engine Code ABA
(2.0 liter/4 cylinders, Motronic)

Engines of vehicles listed below have the engine code **ABA** and **Motronic** engine management system.

Engine displacement: 2.0 liter/4 cylinders. The engine is equipped with the emission influencing components listed in **Table c**.

Model	Model Year
Golf/Jetta	1993, 1994, 1995, 1996, 1997
Cabrio	1994, 1995, 1996, 1997
Passat	1995, 1996, 1997

Table c. Emission Influencing Components: Engine Code ABA

Fuel System	Motronic MFI
Ignition System	Ignition distributor, engine control module (ECM)
Sensors	Mass Air Flow Sensor
	Intake Air Temperature Sensor
	Engine Coolant Temperature Sensor (NTC2)
	Heated Oxygen Sensor
	Throttle Position Sensor
	Knock Sensor
	Camshaft Position Sensor
	RPM Sensor
Control Devices	EVAP canister purge via frequency valve
	Vacuum controlled crankcase ventilation
	Three-Way Catalytic Converter
	Idle speed control via Idle air control valve
	Exhaust gas recirculation (EGR)[1]: EGR Vacuum Regulator Solenoid, EGR Valve, EGR Temp. Sensor
	Diagnostic trouble code MIL Lamp
Other Components	Inlet restrictor

[1] For 50 states, except California vehicles

Emission Diagnosis and Repair

Lambda control loop, checking

Before checking the control loop, be sure that you used the Troubleshooting Chart up to this point. Read Diagnostic Trouble Code (DTC) memory and Output Diagnostic Test Mode (DTM) using the VAG 1551 fault read out unit. If faults are found, correct them first and then clear the fault memory. See applicable Volkswagen Repair Manual.

Test conditions

- Engine oil temperature must be at least 80°C (176°F).
- All electrical consumers such as air conditioner, radio, fan, etc. must be switched off.

1. Connect the following test equipment:
   - Volkswagen fault read out unit, VAG 1551
   - Ignition tester such as VAG 1367
   - Four-gas analyzer, such as VAG 1788 to tailpipe.

2. Run engine at operating temperature.

3. Increase engine speed to 2500 to 2800 rpm for at least 30 seconds until the Lambda value has stabilized.
   - Read first Lambda value and make note of it.

4. Reduce engine speed to idle and wait at least 30 seconds until the Lambda value has stabilized again.
   - Read second Lambda value and make note of it.

5. Remove the vacuum hose at fuel pressure regulator. See Fig. 5-4. Close off hose for 30 seconds and open it again. The Lambda value has to increase briefly and then settle again at the previous value.
   - If the Lambda value did not change, the control loop does not operate properly. In this case reconnect the vacuum hose and continue with the Troubleshooting Chart at "Control loop not OK".
   - If the Lambda value changed, continue checking as follows.

6. After 60 seconds make note of the third Lambda value.

7. Reinstall the vacuum hose on hose connection and secure.

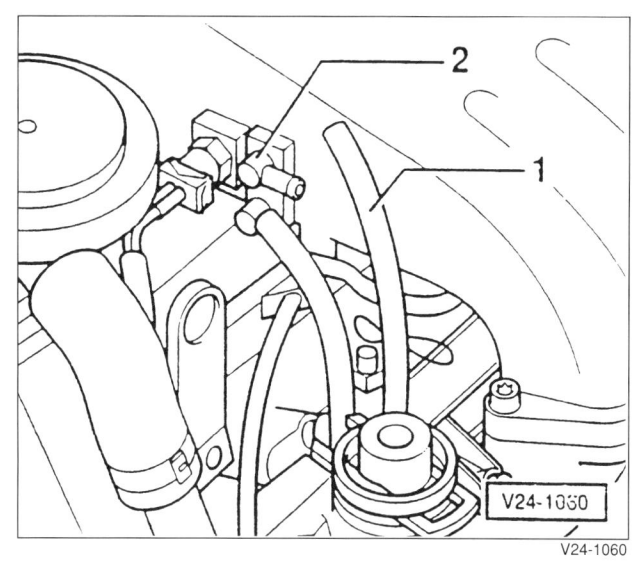

Fig. 5-4. Disconnect vacuum hose (**1**) from hose connection (**2**).

8. Reconnect all hoses or lines that were disconnected.

9. Check whether air conditioner, power steering or automatic transmission affects engine performance.

10. Evaluate noted Lambda values
    - If one or more of the noted Lambda values are out of the range 0.97 to 1.03 or if the Lambda value does not stabilize, the control loop does not function properly. Continue in the Troubleshooting Chart at "Control loop not OK".
    - If the noted Lambda values are within 0.97 to 1.03 and if the Lambda value stabilizes, the control loop is OK. Continue in the Troubleshooting Chart at "Control loop OK".

NOTE

To see the relationship between the Lambda value and oxygen sensor output voltage, see Fig. 2-13 in Chapter 2.

ENGINE CODE ABA (2.0 liter/4 cylinders, Motronic)

Control loop, repairing

Equipment needed:

- Ignition tester such as VAG 1367
- Four Component Tester such as VAG 1788

1. Perform basic engine adjustments as per applicable Volkswagen Repair Manual.

2. If the control loop is still not OK, continue troubleshooting as follows.

Checking and Repairing Emission Influencing Components

Check the following components listed in the table. Perform individual checking steps as per applicable Volkswagen Repair Manual.

NOTE

- *Components that were already recognized as being faulty and were repaired do not have to be rechecked or replaced.*

- *Should you however have found a fault and repaired it, re-check the control loop once again. If the control loop checks out OK, continue in the Troubleshooting Chart with "Control loop OK".*

Additional Components to Test	Volkswagen Repair Manual (Repair manual title and applicable repair group)
Fuel Pressure Regulator	**Fuel Injection and Ignition Repair Group 24**
Throttle Position Switch	
Intake Air Temperature Switch	
Mass Air Flow Sensor	
Fuel Injectors	
Idle Air Control Valve	
Engine Coolant Temperature Sensor	
Knock Sensor	
EGR System	**General, Engine Repair Group 26**
Camshaft Position Sensor	**Fuel Injection and Ignition Repair Group 28**

ENGINE CODE ABA (2.0 liter/4 cylinders, Motronic)

Emission Diagnosis and Repair

Catalytic Converter Check
Vehicles without CO test connection (MY 1993 and later)

Equipment needed:

- Four-gas emission analyzer such as VAG 1788, probe connected to tail pipe.

1. Switch off all electrical consumers such as A/C, radio, fan, lights, etc.

2. Run engine at least for two minutes at 2500 ± 200 rpm to bring catalytic converter to operating temperature.
   - The CO-value shown on the tester should be in the range of 0 to 0.3% of volume.

3. Reduce engine speed to idle.
   - The CO-value shown on the tester should be in the range of 0 to 0.5% of Volume.

4. If one or both CO-values are not within the specified range of 0.3 to 0.5%, continue with Method A described below.

5. If both values are within specifications, continue following the Troubleshooting Chart.

Method A

Equipment needed:

- Four-gas emission analyzer such as VAG 1788 and heat-resistant adapter such as VAG 1363/3

1. Remove union nut for EGR line to exhaust manifold.

2. Remove union nut for EGR line to EGR Valve and remove connector. Swivel EGR line out of the way to make it possible to connect adapter of four-gas analyzer.

3. Retighten union nut at exhaust manifold.

4. Close opening at EGR valve with bolt M16 x 1.5.

5. Connect adapter for four-gas emission analyzer to open end of line.

6. Read CO content (before converter) on read-out of tester at constant rpm and make note of it (A-value in table, Chapter 4).

7. Reconnect EGR line.

8. Check CO-content at tailpipe at same rpm and make note of it (B-value in table, Chapter 4).

9. Calculate percentage using the Troubleshooting Table.
   - Above 50% converter is OK, do not replace.
   - Below 50% converter is not functioning, replace converter.

ENGINE CODE ABA (2.0 liter/4 cylinders, Motronic)

Engine Code ABG
(1.8 liter/4 cylinders, Digifant)

Engines of vehicles listed below have the engine code **ABG** and **Digifant** engine management system.

Engine displacement: 1.8 liter/4 cylinders. The engine is equipped with the emission influencing components listed in **Table d.**

Model	Model Year
Fox	1991, 1992, 1993

Table d. Emission Influencing Components: Engine Code ABG

Fuel system	Digifant II, Digifant I[1]
Ignition System	Ignition distributor, Engine Control Module (ECM)
Sensors	Volume Air Flow Sensor
	Intake Air Temperature Sensor (NTC1)
	Engine Coolant Temperature Sensor (NTC2)
	Heated Oxygen Sensor
	Closed Throttle Position Switch[2], Throttle Position Sensor[1]
	RPM Sensor
Control Devices	Vacuum controlled EVAP canister purge
	Vacuum controlled crankcase ventilation
	Three-Way Catalytic Converter
	Idle speed control via 2 boost valves
	Diagnostic trouble code: Fault memory[1]
Other components	Inlet restrictor

[1] For California vehicles only as of MY 1991
[2] Except California vehicles as of MY 1991

Emission Diagnosis and Repair

Lambda control loop, checking

Before checking the control loop, be sure that you used the Troubleshooting Chart up to this point. Read Diagnostic Trouble Code (DTC) memory and Output Diagnostic Test Mode (DTM) using the VAG 1551 fault read out unit. If faults are found, correct them first and then clear the fault memory. See applicable Volkswagen Repair Manual.

Test conditions

- Engine oil temperature must be at least 80°C (176°F).
- All electrical consumers such as air conditioner, radio, fan, etc. must be switched off.

1. Connect the following test equipment:
   - Volkswagen fault read out unit, VAG 1551
   - Ignition tester such as VAG 1367
   - Four-gas analyzer, such as VAG 1788 to CO test connection on exhaust manifold.

2. Run engine at operating temperature.

3. Increase engine speed to 2500 to 2800 rpm for at least 30 seconds until the Lambda value has stabilized.
   - Read first Lambda value and make note of it.

4. Reduce engine speed to idle and wait at least 30 seconds until Lambda value has stabilized again.
   - Read second Lambda value and make note of it.

5. Evaluate noted Lambda values.
   - If one or more of the noted Lambda values are out of the range 0.97 to 1.03 or if the Lambda value does not stabilize, the control loop does not function properly. Continue in the Troubleshooting Chart at "Control loop not OK".
   - If the noted Lambda values are within 0.97 to 1.03 and if the Lambda value stabilizes, the control loop is OK. Continue in the Troubleshooting Chart at "Control loop OK".

Control loop, repairing

Equipment needed:

- Ignition tester such as VAG 1367 with adapter VAG 1367/8
- CO tester such as VAG 1363 with adapter VAG 1363/3b or SUN105
- Multimeter Fluke 83, US 1119 with adapter 1315 A/2

1. Perform basic engine adjustment as per applicable Volkswagen Repair Manual.
   - Check engine idle speed, adjust if necessary (not adjustable from 7/92)
   - Check CO content, adjust if necessary (not adjustable from 7/92)
   - Check ignition timing, adjust if necessary

2. If the control loop is still not OK, continue troubleshooting as follows.

ENGINE CODE ABG (1.8 liter/4 cylinders, Digifant)

Checking and Repairing Emission Influencing Components

Check the following components listed in the table. Perform individual checking steps as per applicable Volkswagen Repair Manual.

NOTE
- *Components that were already recognized as being faulty and were repaired do not have to be rechecked or replaced.*
- *Should you however have found a fault and repaired it, re-check the control loop once again. If the control loop checks out OK, continue in the Troubleshooting Chart with "Control loop OK".*

Additional Components to Test	Volkswagen Repair Manual (Repair manual title and applicable repair group)
Oxygen sensor	**Fuel Injection and Ignition Repair Group 24**
Volume Air Flow Sensor	
Fuel Pressure Regulator	
Throttle Position Switch	
Throttle switch (only Digifant I)	
Fuel Injectors	
Valve for idle stabilization	
Basic adjustment of throttle	
Ignition timing	**Fuel Injection and Ignition Repair Group 28**
Camshaft Position Sensor	

ENGINE CODE ABG (1.8 liter/4 cylinders, Digifant)

Engine Code 9A
(2.0 liter/4 cylinder (16V), CIS-E Motronic)

Engines of vehicles listed below have the engine code **9A** and **CIS-E Motronic** engine management system.

Engine displacement: 2.0 liter/4 cylinders, 16 valve. The engine is equipped with the emission influencing components listed in **Table e**.

Model	Model Year
Passat 16V	1990, 1991, 1992, 1993
Golf GTI 16V	1990, 1991, 1992
Jetta GLI 16V	1991, 1992

Table e. Emission Influencing Components: Engine Code 9A

Fuel System	CIS-E Motronic
Ignition System	Ignition distributor, Engine Control Module (ECM)
Sensors	Volume Air Flow Sensor
	Engine Coolant Temperature Sensor (NTC2)
	Heated Oxygen Sensor
	Closed Throttle Position Switch
	Wide Open Throttle Position Switch
	2 Knock Sensors
	RPM Sensor
Control Devices	EVAP canister purge via frequency valve
	EVAP canister: located in right front wheel housing
	Vacuum controlled crankcase ventilation
	Three-Way Catalytic Converter
	Idle speed control via Idle Air Control[3], Idle Air Control Valve
	Exhaust gas recirculation (EGR): EGR Vacuum Regulator Solenoid Valve, EGR Valve, EGR Temperature Sensor [1]
	Diagnostic trouble code MIL Lamp[2]
Other components	Inlet restrictor
	Cold Start Injector

[1] EGR for California only
[2] For California only
[3] Up to Model year 1993

66 Chapter 5

Lambda control loop, checking

Before checking the control loop, be sure that you used the Troubleshooting Chart up to this point. Read Diagnostic Trouble Code (DTC) memory and Output Diagnostic Test Mode (DTM) using the VAG 1551 fault read out unit. If faults are found, correct them first and then clear the fault memory. See applicable Volkswagen Repair Manual.

Test conditions

- Engine oil temperature must be at least 80°C (176°F).
- All electrical consumers such as air conditioner, radio, fan, etc. must be switched off.

1. Connect the following test equipment:
   - Volkswagen fault read out unit, VAG 1551
   - Ignition tester such as VAG 1367
   - Four-gas analyzer, such as VAG 1788 to CO test connection on exhaust manifold.

2. Run engine at operating temperature.

3. Increase engine speed to 2500 to 2800 rpm for at least 30 seconds until the Lambda value has stabilized.
   - Read first Lambda value and make note of it.

4. Reduce engine speed to idle and wait at least 30 seconds until Lambda value has stabilized again.
   - Read second Lambda value and make note of it.

5. Reduce engine speed to idle for at least 30 seconds until the Lambda value has stabilized.
   - Read second Lambda value and make note of it.

6. Pull out thin crankcase ventilation hose with connector from hose connection. See Fig. 5-5.

7. Close off connector (**2**).
   - Wait for at least 20 seconds until the Lambda value stabilizes.

8. Open connector (**2**). The Lambda value must increase briefly to stabilize again at the prior value.
   - If the Lambda value does not change, the control loop does not work properly. In this case reconnect the crankcase ventilation hose again and continue in the Troubleshooting Chart at "Control loop not OK".
   - If the Lambda value changes, continue as follows:

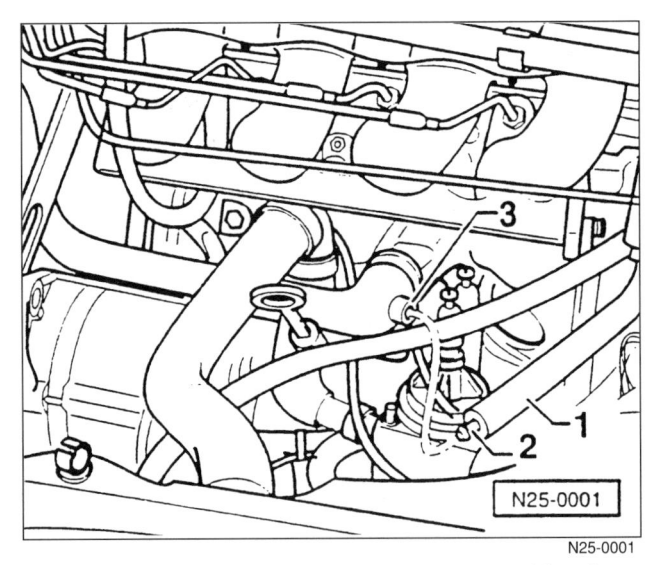

Fig. 5-5. Disconnect crankcase ventilation hose (**1**) at hose connection (**3**).

9. After 60 seconds read the third Lambda value and make note of it.

10. Reconnect all hoses and lines that were disconnected and secure them properly.

11. Check whether running additional components such as A/C or automatic transmission affect engine performance.

12. Evaluate noted Lambda values.
    - If one or more of the noted Lambda values are out of the range 0.97 to 1.03 or if the Lambda value does not stabilize, the control loop does not function properly. Continue in the Troubleshooting Chart at "Control loop not OK".
    - If the noted Lambda values are within 0.97 to 1.03 and if the Lambda value stabilizes, the control loop is OK. Continue in the Troubleshooting Chart at "Control loop OK".

NOTE
To see the relationship between the Lambda value and oxygen sensor output voltage, see Fig. 2-13 in Chapter 2.

ENGINE CODE 9A (2.0 liter/4 cylinder (16V), CIS-E Motronic)

Emission Diagnosis and Repair

Control loop, repairing

Equipment needed:

- Ignition tester such as VAG 1367
- Trigger pliers such as VAG 1367/8
- Adapter such as VAG 1363/3
- CO tester such as VAG 1363 or SUN EPA 75

1. Perform basic engine adjustment as per Volkswagen Repair Manual (Repair Groups 25 and 28).
   - Check engine idle speed (cannot be adjusted).
   - Check CO content, adjust if necessary.
   - Check ignition timing, adjust if necessary.

2. If the control loop is still not OK, continue troubleshooting as follows.

Checking and Repairing Emission Influencing Components

If the control loop is still not OK, check the following components listed in the table. Perform individual checking steps as per applicable Volkswagen Repair Manual.

NOTE
- *Components that were already recognized as being faulty and were repaired do not have to be rechecked or replaced.*

- *Should you however have found a fault and repaired it, re-check the control loop once again. If the control loop checks out OK, continue in the Troubleshooting Chart with "Control loop OK".*

Additional Components to Test	Volkswagen Repair Manual (Repair manual title and applicable repair group)
Oxygen sensor and Lambda regulation	**Fuel Injection and Ignition Repair Group 25**
Volume Air Flow Sensor	
Throttle Position Switch	
Intake Air Temperature Sensor	
Crankshaft Position Sensor	
Idle Air Control Valve	
Cold Start Injector	
Fuel Pressure Regulator	
Fuel Injectors	
Engine Coolant Temperature Sensor	
Full Throttle and Closed Throttle Switches	
Deceleration Fuel Shut-off and Acceleration Enrichment	
EGR System	**General, Engine Repair Group 26**
Camshaft Position Sensor	**Fuel Injection and Ignition Repair Group 28**
Knock Sensors I and II	

ENGINE CODE 9A (2.0 liter/4 cylinder (16V), CIS-E Motronic)

ENGINE CODE: 2H
(1.8 liter/4 cylinder, Digifant)

Engines of vehicles listed below have the engine code **2H** and **Digifant** engine management system.

Engine displacement: 1.8 liter/4 cylinders. The engine is equipped with the emission influencing components listed in **Table f**.

Model	Model Year
Cabriolet	1990, 1991, 1992, 1993

Table f. Emission Influencing Components: Engine Code 2H

Fuel System	Digifant II, Digifant I[1]
Ignition System	Ignition distributor, Engine Control Module (ECM)
Sensors	Volume Air Flow Sensor
	Intake Air Temperature Sensor (NTC1)
	Engine Coolant Temperature Sensor (NTC2)
	Heated Oxygen Sensor
	Idle Throttle Position Switch[2], Wide Open Throttle Position Switch[2], Throttle Position Sensor[1]
	Throttle Position Switch[1]
	Knock Sensor
Control Devices	EVAP canister purge (vacuum controlled)
	Crankcase ventilation (vacuum controlled)
	Three-Way Catalytic Converter
	Automatic Idle speed control
	Diagnostic trouble code: Fault memory[1]
Other Components	Inlet restrictor

[1] For California vehicles only as of MY 1991
[2] Not for California vehicles as of MY 1991

Emission Diagnosis and Repair

Lambda control loop, checking

Before checking the control loop, be sure that you used the Troubleshooting Chart up to this point. Read Diagnostic Trouble Code (DTC) memory and Output Diagnostic Test Mode (DTM) using the VAG 1551 fault read out unit. If faults are found, correct them first and then clear the fault memory. See applicable Volkswagen Repair Manual.

Test conditions

- Engine oil temperature must be at least 80°C (176°F).
- All electrical consumers such as air conditioner, radio, fan, etc. must be switched off.

1. Connect the following test equipment:
   - Volkswagen fault read out unit, VAG 1551
   - Ignition tester such as VAG 1367
   - Four-gas analyzer, such as VAG 1788 to CO test connection on exhaust manifold.

2. Run engine at operating temperature.

3. Increase engine speed to 2500 to 2800 rpm for at least 30 seconds until the Lambda value has stabilized.
   - Read first Lambda value and make note of it.

4. Reduce engine speed to idle and wait at least 30 seconds until Lambda value has stabilized again.

5. Read second Lambda value and make note of it.

6. Remove fuel pressure regulator vacuum hose from connector at intake manifold. See Fig. 5-6. Close connector for 30 seconds and open it again. The Lambda value must increase briefly and then stabilize again at the prior value.
   - If the Lambda value does not change, the control loop does not work properly. Continue in the Troubleshooting Chart at "Control loop not OK".
   - If the Lambda value changes, continue checking as follows.

7. After 60 seconds read the third Lambda value and make note of it.

8. Reconnect all hoses and lines that were disconnected and secure them properly.

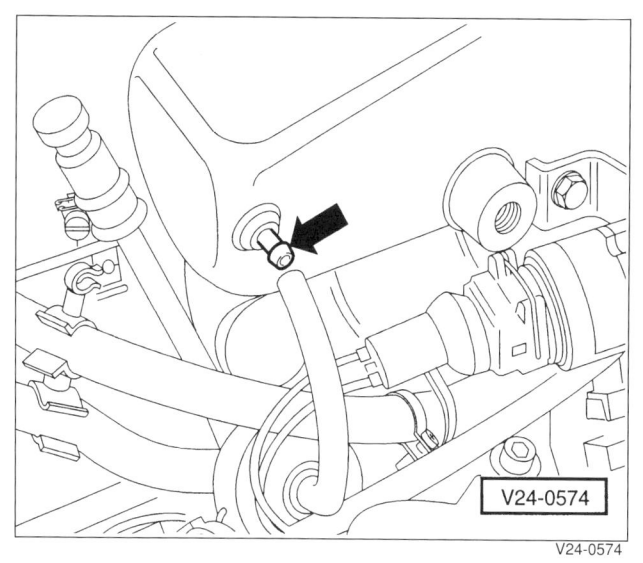

Fig. 5-6. Vacuum hose (arrow) to be disconnected from fuel pressure regulator.

9. Check whether running additional components such as A/C or automatic transmission affect engine performance.

10. Evaluate noted Lambda values.
    - If one or more of the noted Lambda values are out of the range 0.97 to 1.03 or if the Lambda value does not stabilize, the control loop does not function properly. Continue in the Troubleshooting Chart at "Control loop not OK".
    - If the noted Lambda values are within 0.97 to 1.03 and if the Lambda value stabilizes, the control loop is OK. Continue in the Troubleshooting Chart at "Control loop OK".

NOTE

To see the relationship between the Lambda value and oxygen sensor output voltage, see Fig. 2-13 in Chapter 2.

ENGINE CODE: 2H (1.8 liter/4 cylinder, Digifant)

Control loop, repairing

Equipment needed:

- Ignition tester such as VAG 1367 with adapter 1367/8
- CO tester such as VAG 1363 or SUN 105
- Adapter such as VAG 1363/3
- Multimeter Fluke 83, US 1119 with adapter 1315A/2

1. Perform basic engine adjustment as per Volkswagen Repair Manual (Repair Groups 25 and 28).
   - Check engine idle speed (cannot be adjusted).
   - Check CO content, adjust if necessary.

2. If the control loop is still not OK, continue troubleshooting as follows.

Checking and Repairing Emission Influencing Components

Check the following components listed in the table. Perform individual checking steps as per applicable Volkswagen Repair Manual.

NOTE
- *Components that were already recognized as being faulty and were repaired do not have to be rechecked or replaced.*
- *Should you however have found a fault and repaired it, re-check the control loop once again. If the control loop checks out OK, continue in the Troubleshooting Chart with "Control loop OK".*

Additional Components to Test	Volkswagen Repair Manual (Repair manual title and applicable repair group)
Oxygen sensor	**Fuel Injection and Ignition Repair Group 24**
Volume Air Flow Sensor	
Fuel Pressure Regulator	
Throttle Position Switch	
Throttle switch (Digifant I only)	
Fuel Injectors	
Throttle Position Switch (Digifant II only)	
Valves for idle stabilization	
Basic adjustment of throttle	
Ignition timing	
Camshaft Position Sensor	**Fuel Injection and Ignition Repair Group 28**
Knock Sensor	

ENGINE CODE: 2H (1.8 liter/4 cylinder, Digifant)

Engine Code CV (2.0 liter/4 cylinder, AFC (L-Jetronic)

Engines of vehicles listed below have the engine code **CV** and **AFC (L-Jetronic)** engine management system.

Engine displacement: 2.0 liter/4 cylinders. The engine is equipped with the emission influencing components listed in **Table g**.

Model	Model Year
Vanagon (air-cooled)	1980, 1981, 1982, 1983

Table g. Emission Influencing Components: Engine Code CV

Fuel System	AFC
Ignition System	Ignition distributor, Vacuum Diaphragm
Sensors	Volume Air Flow Sensor
	Intake Air Temperature Sensor (NTC1)
	Engine Coolant Temperature Sensor (NTC2)
	Oxygen Sensor[2]
	Closed Throttle Position Switch
	Wide Open Throttle Switch
Control Devices	Vacuum controlled EVAP canister purge
	Vacuum controlled crankcase ventilation
	Three-Way Catalytic Converter/Oxidation catalyst[1]
	Idle speed control: Digital Idle Stabilization (DIS)[2]
	Exhaust gas recirculation (EGR): EGR Valve (not for California)
	Diagnostic Trouble Code: MIL lamp[2]
Other components	Solenoid Starting Valve
	Deceleration Valve[3]

[1] Federal certification with oxidation catalytic converter and without oxygen sensor
[2] For California only
[3] For manual transmissions only

Lambda control loop, checking

Before checking the control loop, be sure that you used the Troubleshooting Chart up to this point.

A. For California vehicles only

Test conditions

- Engine oil temperature must be at least 80°C (176°F).
- All electrical consumers such as air conditioner, radio, fan, etc. must be switched off.

1. Connect the following test equipment:
   - Ignition tester such as VAG 1367
   - Four-gas analyzer, such as VAG 1788 to CO test connection on exhaust manifold.

2. Run engine at operating temperature.

3. Increase engine speed to 2500 to 2800 rpm for at least 30 seconds until the Lambda value has stabilized.
   - Read first Lambda value and make note of it.

4. Reduce engine speed to idle and wait at least 30 seconds until Lambda value has stabilized again.

5. Read second Lambda value and make note of it.

6. Evaluate noted Lambda values.
   - If one or more of the noted Lambda values are out of the range 0.97 to 1.03 or if the Lambda value does not stabilize, the control loop does not function properly. Continue in the Troubleshooting Chart at "Control loop not OK".
   - If the noted Lambda values are within 0.97 to 1.03 and if the Lambda value stabilizes, the control loop is OK. Continue in the Troubleshooting Chart at "Control loop OK".

NOTE

To see the relationship between the Lambda value and oxygen sensor output voltage, see Fig. 2-13 in Chapter 2.

Control loop, repairing

Equipment needed:

- Ignition tester such as VAG 1367 with adapter 1473
- CO tester such as VAG 1363 or SUN 105
- Multimeter Fluke 83, US 1119 with adapter 1315A/2

1. Perform basic engine adjustment as per Volkswagen Repair Manual (Repair Groups 25 and 28).
   - Check engine idle speed (cannot be adjusted).
   - Check CO content, adjust if necessary.

2. If the control loop is still not OK, continue troubleshooting as follows.

ENGINE CODE CV (2.0 liter/4 cylinder, AFC (L-Jetronic)

Emission Diagnosis and Repair

Checking and Repairing Emission Influencing Components

Check the following components listed in the table. Perform individual checking steps as per applicable Volkswagen Repair Manual.

NOTE
- *Components that were already recognized as being faulty and were repaired do not have to be rechecked or replaced.*
- *Should you however have found a fault and repaired it, re-check the control loop once again. If the control loop checks out OK, continue in the Troubleshooting Chart with "Control loop OK".*

Check and repair the following emission influencing components:

Additional Components to Test	Volkswagen Repair Manual (Repair manual title and applicable repair group)
Oxygen sensor	**Fuel Injection and Ignition Repair Group 24**
Electrical System	
Volume Air Flow Sensor	
Intake Air Temperature Sensor	
Engine Coolant Temperature Sensor	
Thermal Time Switch	
Fuel Pressure Regulator	
Fuel Injectors	
Throttle Position Switch	
Deceleration Valve (manual transmission only)	
Cold Start Injector	
Ignition timing	**Fuel Injection and Ignition Repair Group 28**
Camshaft Position Sensor	

ENGINE CODE CV (2.0 liter/4 cylinder, AFC (L-Jetronic)

Engine Code DH
(1.9 liter/4 cylinder, Digijet)

Engines of vehicles listed below have the engine code **DH** and **Digijet** engine management system. Engine displacement: 1.9 liter/4 cylinders.

The engine is equipped with the emission influencing components listed in **Table h**.

Model	Model Year
Vanagon (water-cooled)	1983, 1984, 1985

Table h. Emission Influencing Components: Engine Code DH

Fuel System	Digijet-MFI
Ignition System	Ignition distributor, Vacuum Diaphragm, ECM[1]
Sensors	Volume Air Flow Sensor
	Intake Air Temperature Sensor (NTC1)
	Engine Coolant Temperature Sensor (NTC2)
	Oxygen Sensor
	Closed Throttle Position Switch
	Wide Open Throttle Position Switch
Control Devices	Vacuum controlled EVAP canister purge
	Vacuum controlled crankcase ventilation
	Three-Way Catalytic Converter
	Idle speed control: Digital Idle Stabilization[2]
	Exhaust gas recirculation (EGR): EGR Valve[2]
Other Components	Auxiliary Air Valve

[1] Timing advance controlled by vacuum diaphragm, idle speed controlled by ECM
[2] For California only

Emission Diagnosis and Repair

Lambda control loop, checking

Before starting the control loop be sure that you used the Troubleshooting Chart up to this point.

Test conditions

- Engine oil temperature must be at least 80°C (176 °F).
- All electrical consumers such as air conditioner, radio, fan, etc. must be switched off.

1. Connect the following test equipment:
   - Ignition tester such as VAG 1367
   - Four-gas emission analyzer, such as VAG 1788 connected to CO test connector.

2. Increase engine speed to 2500 to 2800 rpm for at least 30 seconds until the Lambda value has stabilized.
   - Read first Lambda value and make note of it.

3. Reduce engine speed to idle for at least 30 seconds until the Lambda value has stabilized again.
   - Read second Lambda value and make note of it.

4. Remove vacuum hose from fuel pressure regulator. See Fig. 5-7. Close off hose for at least 30 seconds, then open it again. The Lambda value must increase briefly and then stabilize again at the prior value.
   - If the Lambda value does not change, the control loop does not work properly. Continue in the Troubleshooting Chart at "Control loop not OK"
   - If the Lambda value changes continue as follows.

5. After 60 seconds read the third Lambda value and make note of it.

6. Reconnect all hoses and lines that were disconnected and secure them properly.

7. Check whether running additional components such as A/C or automatic transmission affect engine performance.

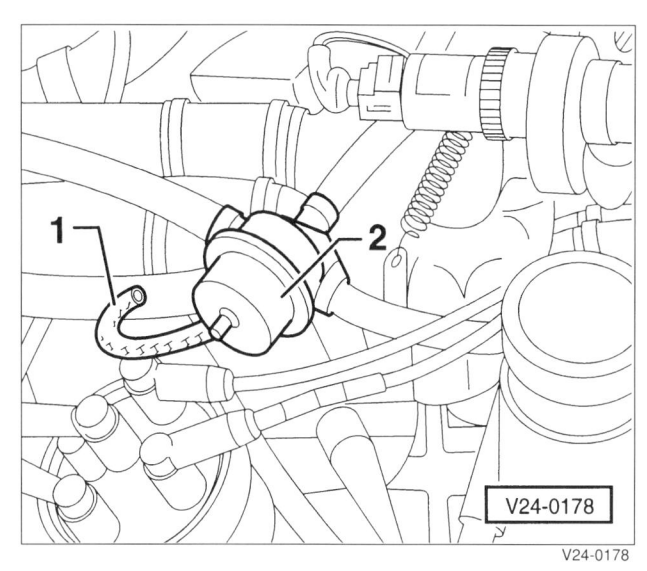

Fig. 5-7. Vacuum hose (arrow) to be disconnected from fuel pressure regulator.

8. Evaluate noted Lambda values.
   - If one or more of the noted Lambda values are out of the range 0.97 to 1.03 or if the Lambda value does not stabilize, the control loop does not function properly. Continue in the Troubleshooting Chart at "Control loop not OK".
   - If the noted Lambda values are within 0.97 to 1.03 and if the Lambda value stabilizes, the control loop is OK. Continue in the Troubleshooting Chart at "Control loop OK".

NOTE

To see the relationship between the Lambda value and oxygen sensor output voltage, see Fig. 2-13 in Chapter 2.

ENGINE CODE DH (1.9 liter/4 cylinder, Digijet)

Control loop, repairing

Equipment needed:

- Ignition tester such as VAG 1367 with adapter 1473
- CO tester such as VAG 1363 or SUN 105
- Multimeter Fluke 83, US 1119 with adapter 1315A/2

1. Perform basic engine adjustment as per applicable Volkswagen Repair Manual (Repair Groups 24 and 28).
   - Check engine idle speed, adjust if necessary.
   - Check CO content, adjust if necessary.

2. If the control loop is still not OK, continue troubleshooting as follows.

Checking and Repairing Emission Influencing Components

Check the following components listed in the table. Perform individual checking steps as per applicable Volkswagen Repair Manual.

NOTE
- *Components that were already recognized as being faulty and were repaired do not have to be rechecked or replaced.*
- *Should you however have found a fault and repaired it, re-check the control loop once again. If the control loop checks out OK, continue in the Troubleshooting Chart with "Control loop OK".*

Additional Components to Test	Volkswagen Repair Manual (Repair manual title and applicable repair group)
Power supply for Engine Control Module (ECM)	**Fuel Injection and Ignition Repair Group 24**
Ignition Control Module	
Fuel Injectors	
Temperature Sensor II	
Wide Open Throttle Switch (F25 or F60/F81)	
Volume Air Flow Sensor Position Sensor	
Intake Air Temperature Sensor	
Wiring to Heated Oxygen Sensor	
Volume Air Flow Sensor	
Oxygen Sensor	
Fuel Pressure Regulator	
Fuel Injectors	
Throttle Position Switch	
Idle Air Control valve	
Ignition timing	**Fuel Injection and Ignition Repair Group 28**
Camshaft Position Sensor	

Emission Diagnosis and Repair

ENGINE CODE EJ
(1.6 liter/4 cylinder, CIS)

Engines of vehicles listed below have the engine code **EJ** and **CIS** engine management system.

Engine displacement: 1.6 liter/4 cylinders. The engine is equipped with the emission influencing components listed in **Table j**.

Model	Model Year
Rabbit	1980
Rabbit Convertible	1980
Scirocco	1980
Pick-up	1980

Table j. Emission Influencing Components: Engine Code EJ

Fuel System	CIS / CIS with and without Lambda regulation
Ignition System	Ignition distributor
Sensors	Volume Air Flow Sensor
	Cooling Fan Switch
	Oxygen Sensor[2]
Control Devices	Vacuum controlled EVAP canister purge
	EVAP canister: located in right front wheel housing
	Vacuum controlled crankcase ventilation
	Oxidation Catalytic Converter[3], Three-Way Catalytic Converter[2], None[4]
	Idle speed control: Digital Idle Stabilization[2]
	Exhaust gas recirculation (EGR): Thermal Time Switch and EGR Valve[4]
Other Components	Solenoid Starting Valve[1]
	Warm-up Unit

[1] Pick-up only
[2] Rabbit, Cabrio and Scirocco, California models
[3] Pick-Up, California model
[4] Pick-Up, 50 States, except California

Engine Code EN
(1.7 liter/4 cylinder, CIS)

Engines of vehicles listed below have the engine code **EN** and **CIS** engine management system.

Engine displacement: 1.7 liter/4 cylinders. The engine is equipped with the emission influencing components listed in **Table i**.

Model	Model Year
Rabbit	1981, 1982, 1983, 1984
Rabbit Convertible	1981, 1982, 1983
Jetta	1981, 1982, 1983, 1984
Scirocco	1981, 1982, 1983
Quantum	1983
Pick-up	1981, 1982, 1983

Table i. Emission Influencing Components: Engine Code EN

Fuel System	CIS / CIS with Lambda regulation[1]
Ignition System	Ignition distributor, Vacuum Diaphragm, ECM[3]
Sensors	Volume Air Flow Sensor
	Cooling Fan Switch
	Oxygen Sensor[1]
Control Devices	Vacuum controlled EVAP canister purge
	EVAP canister: located in right front wheel housing[5]
	Vacuum controlled crankcase ventilation
	Three-Way Catalytic Converter
	Idle speed control: Solenoid Air Valve[2] / Digital Idle Stabilization[3]
	Exhaust gas recirculation (EGR): Thermal Time Switch and EGR Valve[4]
Other Components	Solenoid Starting Valve[4]
	Frequency Valve[4]
	2 EGR Control Valves[4]
	Warm-up Unit

[1] Pick-up, model years 1981–1983 for 50 States (not California) with oxidation catalytic converter and without oxygen sensor
[2] Model year 1984 only
[3] Pick-up for 50 states (not California) without idle speed control
[4] With oxidation catalytic converter
[5] Not known for Convertible, Jetta, Scirocco, model year 1982

ENGINE CODE EJ/EN

Lambda control loop, checking (vehicles with oxygen sensor)

Before starting the control loop be sure that you used the Troubleshooting Chart up to this point.

Test conditions

- Engine oil temperature must be at least 80°C (176°F).
- All electrical consumers such as air conditioner, radio, fan, etc. must be switched off.

1. Connect the following test equipment:
   - Ignition tester such as VAG 1367
   - Four-gas emission analyzer, such as VAG 1788 connected to CO test connector.

2. Increase engine speed to 2500 to 2800 rpm for at least 30 seconds until the Lambda value has stabilized.
   - Read first Lambda value and make note of it.

3. Reduce engine speed to idle for at least 30 seconds until the Lambda value has stabilized again.
   - Read second Lambda value and make note of it.

4. Evaluate noted Lambda values.
   - If one or more of the noted Lambda values are out of the range 0.97 to 1.03 or if the Lambda value does not stabilize, the control loop does not function properly. Continue in the Troubleshooting Chart at "Control loop not OK".
   - If the noted Lambda values are within 0.97 to 1.03 and if the Lambda value stabilizes, the control loop is OK. Continue in the Troubleshooting Chart at "Control loop OK".

NOTE

To see the relationship between the Lambda value and oxygen sensor output voltage, see Fig. 2-13 in Chapter 2.

5. Remove test hose from CO test connector and close connector.

6. Check whether running additional components such as A/C or automatic transmission affect engine performance.

Control loop, repairing

Equipment needed:

- Ignition tester such as Siemens 451 or SUN TDT-11
- Adapter such as US1112
- CO tester such as VAG 1363 or SUN EPA 75, connected to CO test connector

1. Perform basic engine adjustments as per applicable Volkswagen Repair Manual (Repair Groups 25 and 28).
   - Check engine idle speed, adjust f necessary
   - Check ignition timing, adjust if necessary.
   - Check CO-content, adjust if necessary.

2. If the control loop is still not OK, continue troubleshooting as follows.

Checking and Repairing Emission Influencing Components

Check the following components in the table. Perform individual checking steps as per applicable Volkswagen Repair Manual.

NOTE
- *Components that were already recognized as being faulty and were repaired do not have to be rechecked or replaced.*
- *Should you however have found a fault and repaired it, re-check the control loop once again. If the control loop checks out OK, continue in the Troubleshooting Chart with "Control loop OK".*
- *For vehicles without Lambda regulation, check after every repair that the basic engine adjustment is OK. If the engine is within specifications according to technical data in the repair manual, continue in the Troubleshooting Chart at "Engine adjustment OK"*

Additional Components to Test	Volkswagen Repair Manual (Repair manual title and applicable repair group)
Air Sensor Plate	**Fuel Injection and Ignition Repair Group 25**
Fuel Pressure Regulator	
Frequency valve	
Idle Air Control Valve	
Auxiliary Air Regulator	
Cold Start Injector	
Thermal Time Switch	
Fuel Injectors	
Deceleration Air Valve	**General, Engine Repair Group 26**
Components of the control loop	
EGR Valve and EGR Thermal Valve (only for Pick-Up with "Federal" certification and oxidation catalytic converter)	
Ignition related problems	**Fuel Injection and Ignition Repair Group 28**

Lambda control loop, checking (vehicles without oxygen sensor)

Equipment needed:
- Ignition tester such as Siemens 451 or SUN TDT-11
- Adapter such as US1112
- CO tester such as VAG 1363 or SUN EPA 75, connected to CO test connector.

1. Perform basic engine adjustment as per Volkswagen Repair Manual (Repair Groups 25 and 28).
   - Check engine idle speed, adjust if necessary
   - Check ignition timing, adjust if necessary.
   - Check CO content, adjust if necessary.

2. If basic engine adjustments cannot be made correctly, the engine must be repaired. Continue with "Check and repair emission influencing components," given earlier.

Engine Code GX
(1.8 liter/4 cylinder, CIS/CIS-E)

Engines of vehicles listed below have the engine code **GX** and **CIS or CIS-E** engine management system.

Engine displacement: 1.8 liter/4 cylinders. The engine is equipped with the emission influencing components listed in **Table k.**

Model	Model Year
Golf	1985, 1986, 1987
Jetta	1985, 1986, 1987

Table k. Emission Influencing Components: Engine Code GX

Fuel System	CIS / CIS-E[1]
Ignition System	Ignition distributor, Vacuum Diaphragm
Sensors	Volume Air Flow Sensor
	Engine Coolant Temperature Sensor (NTC 2)[2]
	Cooling Fan Switch[2]
	Oxygen Sensor
Control Devices	EVAP canister purge (vacuum controlled)
	EVAP canister: located in right front wheel housing
	Crankcase ventilation (vacuum controlled)
	Three-Way Catalytic Converter
	Idle speed control: Idle Air Control Valve
Other Components	Solenoid Starting Valve
	Differential pressure regulator/Frequency Valve[3]
	Warm-up Unit[1]

[1] CIS as well as CIS-E was installed in engines with engine code GX. The difference is the warm-up unit (control pressure regulator) for CIS was replaced with the differential pressure regulator for CIS-E
[2] Cooling fan switch for CIS
[3] Differential pressure regulator for CIS-E, Frequency valve for CIS

82 Chapter 5

Lambda control loop, checking

Before starting the control loop be sure that you used the Troubleshooting Chart up to this point.

Test conditions

- Engine oil temperature must be at least 80°C (176°F).
- All electrical consumers such as air conditioner, radio, fan, etc. must be switched off.
- Hose from two-way solenoid valve for idle speed boost must be pinched (closed).
- Vehicles with automatic transmission: accelerator cable must be properly adjusted.

1. Connect the following test equipment:
   - Ignition tester such as VAG 1367
   - Four-gas emission analyzer, such as VAG 1788 connected to CO test connector.

2. Run engine at idle (at operating temperature)

3. Increase engine speed to 2500 to 2800 rpm for at least 30 seconds until the Lambda value has stabilized.
   - Read first Lambda value and make note of it.

4. Reduce engine speed to idle and wait at least 30 seconds until the Lambda value has stabilized again.
   - Read second Lambda value and make note of it.

5. Remove crankcase venational hose from connection on intake manifold. See Fig. 5-8. Close off hose connection on intake manifold and open again after 30 seconds. The Lambda value must increase briefly and then stabilize again at the prior value.
   - If the Lambda value does not change, the control loop does not work properly. Continue in the Troubleshooting Chart at "Control loop not OK". In this case reconnect the disconnected hose.
   - If the Lambda value changes continue checking as follows.

6. After 60 seconds read the third Lambda value and make note of it.

7. Reconnect crankcase ventilation hose.

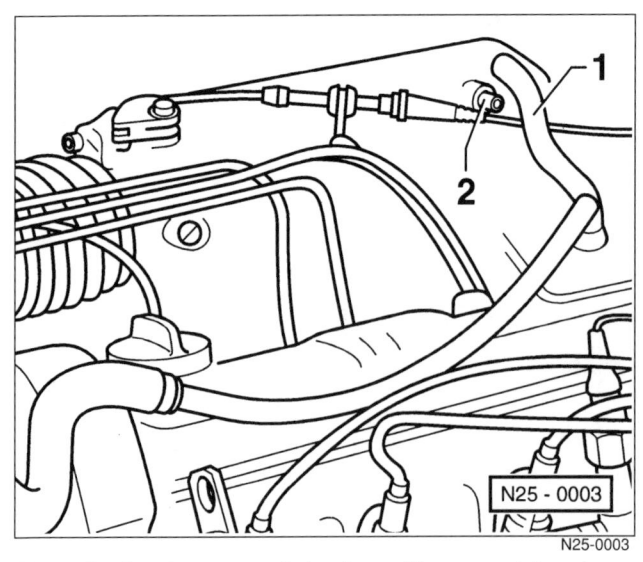

Fig. 5-8. Crankcase ventilation hose (**1**) removed from hose connection (**2**).

8. Reconnect all hoses and lines that were disconnected and secure them properly.

9. Check whether running additional components such as A/C or automatic transmission affect engine performance.

10. Evaluate noted Lambda values.
    - If one or more of the noted Lambda values are out of the range 0.97 to 1.03 or if the Lambda value does not stabilize, the control loop does not function properly. Continue in the Troubleshooting Chart at "Control loop not OK".
    - If the noted Lambda values are within 0.97 to 1.03 and if the Lambda value stabilizes, the control loop is OK. Continue in the Troubleshooting Chart at "Control loop OK".

NOTE

To see the relationship between the Lambda value and oxygen sensor output voltage, see Fig. 2-13 in Chapter 2.

ENGINE CODE GX (1.8 liter/4 cylinder, CIS/CIS-E)

Emission Diagnosis and Repair

Control loop, repairing

Equipment needed:

- Ignition tester such as VAG 1367
- Trigger Pliers such as US 1112
- Adapter such as VAG 1363/3
- CO tester such as VAG 1363 or SUN EPA 75
- Multimeter Fluke 83, US 1119

1. Perform basic engine adjustment as per Volkswagen Repair Manual (Repair Groups 25 and 28).
   - Check engine idle speed, adjust if necessary.
   - Check CO content, adjust if necessary.
   - Check ignition timing, adjust if necessary.

2. If the control loop is still not OK, continue troubleshooting.

Checking and Repairing Emission Influencing Components

Check the following components in the table. Perform individual checking steps as per applicable Volkswagen Repair Manual.

NOTE
- *Components that were already recognized as being faulty and were repaired do not have to be rechecked or replaced.*

- *Should you however have found a fault and repaired it, re-check the control loop once again. If the control loop checks out OK, continue in the Troubleshooting Chart with "Control loop OK".*

- *Components that apply specifically to either CIS or CIS-E are identified below. Components that apply to CIS as well as to CIS-E are listed without further identification.*

Additional Components to Test	Volkswagen Repair Manual (Repair manual title and applicable repair group)
Differential Fuel Pressure Regulator (CIS-E only)	**Fuel Injection and Ignition Repair Group 25**
Engine Coolant Temperature Sensor	
Power supply for Engine Control Module (J21), Fuse S5, and Power Supply Relay (J16)	
Wiring to Thermal Time Switch (F58) (CIS only)	
Fuse, Relay 26 (J169, Frequency Valve (N7), connector for duty cycle (CIS only)	
Oxygen Sensor and Lambda regulation	
Air Sensor Plate	
Volume Air Flow Position Sensor (CIS-E only)	
Cold Start Injector	
Thermal Time Switch	
Fuel Pressure Regulator	
Differential pressure regulator (CIS-E only)	
Fuel Injectors	
Idle Air Control Valve (N62 I and II) for idle boost (CIS-E)	
Ignition related problems	**Fuel Injection and Ignition Repair Group 28**

ENGINE CODE GX (1.8 liter/4 cylinder, CIS/CIS-E)

Engine Code HT (1.8 liter/4 cylinder, CIS-E)

Engines of vehicles listed below have the engine code **HT** and **CIS-E** engine management system.

Engine displacement: 1.8 liter/4 cylinders. The engine is equipped with the emission influencing components listed in **Table I**.

Model	Model Year
Golf GTI	1985, 1986
Jetta GLI	1985, 1986

Table I. Emission Influencing Components: Engine Code HT

Fuel System	CIS-E
Ignition System	Ignition distributor, ECM
Sensors	Volume Air Flow Sensor
	Engine Coolant Temperature Sensor (NTC 2)
	Heated Oxygen Sensor
	Closed Throttle Switch
	Wide Open Throttle Switch
	Knock Sensor
Control Devices	Vacuum controlled EVAP canister purge
	Vacuum controlled crankcase ventilation
	Three-Way Catalytic Converter
	Idle speed control: Idle Air Control Valve
	Diagnostic Trouble Codes (DTC): Ignition System
Other Components	Solenoid Starting Valve
	Electric Hydraulic Regulator

Emission Diagnosis and Repair 85

Lambda control loop, checking

Before starting the control loop be sure that you used the Troubleshooting Chart up to this point.

Test conditions

- Engine oil temperature must be at least 80°C (176 °F).
- All electrical consumers such as air conditioner, radio, fan, etc. must be switched off.

1. Connect the following test equipment:
   - Ignition tester such as VAG 1367
   - Four-gas emission analyzer, such as VAG 1788 connected to CO test connector.

2. Increase engine speed to 2500 to 2800 rpm for at least 30 seconds until the Lambda value has stabilized.
   - Read first Lambda value and make note of it.

3. Reduce engine speed to idle for at least 30 seconds until the Lambda value has stabilized again.
   - Read second Lambda value and make note of it.

4. Remove crankcase venational hose from connection on intake manifold. See Fig. 5-9. Close off hose connection on intake manifold and open again after 30 seconds. The Lambda value must increase briefly and then stabilize again at the prior value.
   - If the Lambda value does not change, the control loop does not work properly. Continue in the Troubleshooting Chart at "Control loop not OK". In this case reconnect the disconnected hose.
   - If the Lambda value changes continue checking as follows.

5. After 60 seconds read the third Lambda value and make note of it.

6. Reconnect disconnected hose. Reconnect all hoses and lines that were disconnected and secure them properly.

7. Check whether running additional components such as A/C or automatic transmission affects engine performance.

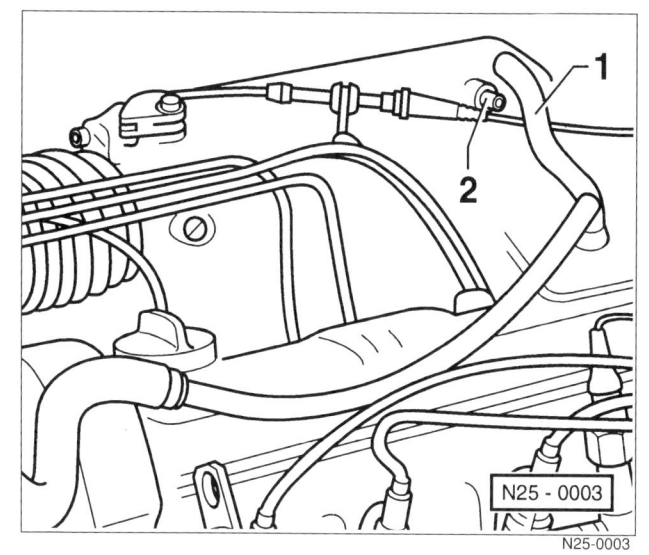

Fig. 5-9. Crankcase ventilation hose (**1**) removed from hose connection (**2**).

8. Evaluate noted Lambda values.
   - If one or more of the noted Lambda values are out of the range 0.97 to 1.03 or if the Lambda value does not stabilize, the control loop does not function properly. Continue in the Troubleshooting Chart at "Control loop not OK".
   - If the noted Lambda values are within 0.97 to 1.03 and if the Lambda value stabilizes, the control loop is OK. Continue in the Troubleshooting Chart at "Control loop OK".

Control loop, repairing

Equipment needed:

- Ignition tester such as VAG 1367
- Trigger Pliers such as US 1112
- Adapter such as VAG 1363/3
- CO tester such as VAG 1363 or SUN 105
- Multimeter Fluke 83, US 1119
- Adapter VAG 1315 A/1

1. Perform basic engine adjustment as per Volkswagen Repair Manual (Repair Groups 25 and 28).
   - Check engine idle speed, adjust if necessary.
   - Check CO content, adjust if necessary.

2. If the control loop is still not OK, continue in the troubleshooting.

ENGINE CODE HT (1.8 liter/4 cylinder, CIS-E)

Checking and Repairing Emission Influencing Components

Check the following individual components as per Volkswagen Repair Manual (Repair Group 25 and Group 28).

NOTE
- *Components that were already recognized as being faulty and were repaired do not have to be rechecked or replaced.*
- *Should you however have found a fault and repaired it, re-check the control loop once again. If the control loop checks out OK, continue in the Troubleshooting Chart with "Control loop OK".*

- Heated Oxygen Sensor (G39)
- Idle Air Control Valve (N71)
- Volume Air Flow Sensor Position Sensor (G19)
- Engine Coolant Temperature Sensor (N10)
- Power supply for Ignition Control Module (J154)
- Wide Open Throttle Switch (F81)
- Closed Throttle Position Switch (F60)
- Signal from Fuel Pump Relay (J17)
- Wiring to Engine Control Module (ECM) (J21), pin 25
- Wiring for On Board Diagnostics
- Wiring to Ignition Control Module (N41), pin 6
- Wiring to Camshaft Position Sensor (G40)
- Knock sensor (G61)

If the control loop is still not OK, continue with checking the following components:

Additional Components to Test	Volkswagen Repair Manual (Repair manual title and applicable repair group)
Oxygen Sensor and Lambda regulation	**Fuel Injection and Ignition Repair Group 25**
Fuel Injectors	
Closed Throttle Position Switch (F60) and Wide Open Throttle Witch (F81)	
Air Sensor Plate	
Volume Air Flow Position Sensor	
Cold Start Injector	
Cold Start Injector Thermal Switch (F26)	
Fuel Pressure Regulator	
Differential Pressure Regulator	
Idle Air Control Valve (N62 I and II) of idle boost	
Ignition Timing	**Fuel Injection and Ignition Repair Group 28**

ENGINE CODE HT (1.8 liter/4 cylinder, CIS-E)

Engine Code JH
(1.8 liter/4 cylinder, CIS)

Engines of vehicles listed below have the engine code **JH** and **CIS with Lambda Regulation** engine management system.

Engine displacement: 1.8 liter/4 cylinders. The engine is equipped with the emission influencing components listed in **Table m**.

Model	Model Year
Scirocco	1983, 1984, 1985, 1986, 1987
Cabriolet	1985, 1986, 1987, 1988, 1989
Rabbit GTI	1983, 1984
Rabbit Convertible	1984
Jetta	1984

Table m. Emission Influencing Components: Engine Code JH

Fuel System	CIS with Lambda regulation
Ignition System	Ignition distributor, vacuum diaphragm
Sensors	Volume Air Flow Sensor
	Cooling Fan Switch
	Oxygen Sensor
	Wide Open Throttle Position Switch
Control Devices	Vacuum controlled EVAP canister purge
	EVAP canister: located in right front wheel housing
	Vacuum controlled crankcase ventilation
	Three-Way Catalytic Converter
	2 boost valves[2]
Other Components	Solenoid Starting Valve
	Frequency Valve
	Warm-up Unit[1]

[1] From MY 1983–1987
[2] Not M Y 1983

Chapter 5

Lambda control loop, checking

Before starting to check the control loop be sure that you used the Troubleshooting Chart up to this point.

Test conditions

- Engine oil temperature must be at least 80°C (176 °F).
- All electrical consumers such as air conditioner, radio, fan, etc. must be switched off.
- Without idle speed boost: Switch on headlights.
- With idle speed boost: Pinch (close) hose from two-way valve for idle speed boost.

1. Connect the following test equipment:
   - Ignition tester such as VAG 1367
   - Four-gas emission analyzer, such as VAG 1788 connected to CO test connector.

2. Run engine at idle.

3. Increase engine speed to 2500 to 2800 rpm for at least 30 seconds until the Lambda value has stabilized.
   - Read first Lambda value and make note of it.

4. Reduce engine speed to idle and wait at least 30 seconds until the Lambda value has stabilized again.
   - Read second Lambda value and make note of it.

5. Remove hose or cap from vacuum line (see Fig. 5-10). The Lambda value must increase briefly and then stabilize again at the prior value.
   - If the Lambda value does not change, the control loop does not work properly. Continue in the Troubleshooting Chart at "Control loop not OK". In this case reconnect the hose or cap.
   - If the Lambda value changes continue as follows.

6. After 60 seconds read the third Lambda value and make note of it.

7. Reconnect disconnected hose or cap.

8. Reconnect all hoses and lines that were disconnected and secure them properly.

9. Check whether running additional components such as A/C or automatic transmission affect engine performance.

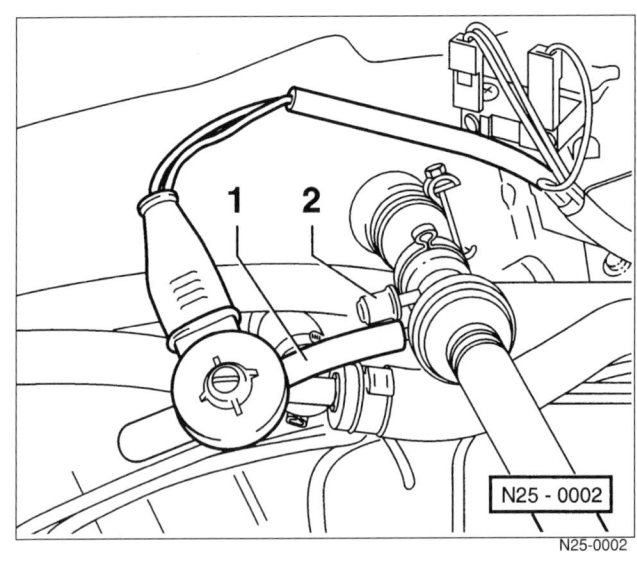

Fig. 5-10. Vacuum line hose (1) or cap (2) to be disconnected from large vacuum line.

10. Evaluate noted Lambda values.
    - If one or more of the noted Lambda values are out of the range 0.97 to 1.03 or if the Lambda value does not stabilize, the control loop does not function properly. Continue in the Troubleshooting Chart at "Control loop not OK".
    - If the noted Lambda values are within 0.97 to 1.03 and if the Lambda value stabilizes, the control loop is OK. Continue in the Troubleshooting Chart at "Control loop OK".

NOTE

To see the relationship between the Lambda value and oxygen sensor output voltage, see Fig. 2-13 in Chapter 2.

ENGINE CODE JH (1.8 liter/4 cylinder, CIS)

Emission Diagnosis and Repair

Control loop, repairing

Equipment needed:

- Ignition tester such as VAG 1367
- Trigger pliers such as VAG 1367/8
- Adapter such as VAG 1363/3
- CO tester such as VAG 1363 or SUN EPA 75

1. Perform basic engine adjustment as per Volkswagen Repair Manual (Repair Groups 24 and 28).
   - Check engine idle speed, adjust if necessary.
   - Check CO content, adjust if necessary.

2. If the control loop is still not OK, continue with the troubleshooting.

Checking and Repairing Emission Influencing Components

Check the following components in the table. Perform individual checking steps as per applicable Volkswagen Repair Manual.

NOTE
- *Components that were already recognized as being faulty and were repaired do not have to be rechecked or replaced.*
- *Should you however have found a fault and repaired it, re-check the control loop once again. If the control loop checks out OK, continue in the Troubleshooting Chart with "Control loop OK".*

Additional Components to Test	Volkswagen Repair Manual (Repair manual title and applicable repair group)
Oxygen Sensor	**Fuel Injection and Ignition, Repair Group 25**
Secondary Air Valve	
Cold Start Injector and Thermal Time Switch	
Fuel pressure	
Fuel Injectors	
Wide Open Throttle Position Switch	
Air Sensor Plate, position	
Air Sensor Plate, rest position and free play	
Differential Pressure Regulator	
Ignition system components	**Fuel Injection and Ignition, Repair Group 28**

ENGINE CODE JH (1.8 liter/4 cylinder, CIS)

ENGINE CODE UM/JN (1.8 liter/4 cylinder, CIS-E)

Engines of vehicles listed below have the engine code **UM/JN** and **CIS-E** engine management system. In model year 1990, the engine code UM was changed to JN.

Engine displacement: 1.8 liters/4 cylinders. The engine is equipped with the emission influencing components listed in **Table n**.

Model	Model Year
Fox/Wagon	1987, 1988, 1989, 1990*
Quantum/Wagon	1984, 1985, 1986

Table n. Emission Influencing Components: Engine Code UM/JN

Fuel system	CIS-E
Ignition System	Vacuum Diaphragm, Ignition distributor, Engine Control Module (ECM)
Sensors	Volume Air Flow Sensor
	Engine Coolant Temperature Sensor (NTC2)
	Heated Oxygen Sensor
	Closed Throttle Position Switch
Control Devices	EVAP canister purge via frequency valve
	Vacuum controlled crankcase ventilation
	Three-Way Catalytic Converter
	Vacuum controlled Exhaust Gas Recirculation (EGR)[1]
	Idle speed control via 2 boost valves
	DTC: Ignition System
Other components	Inlet restrictor
	Starting valve

[1] For California vehicles only

Emission Diagnosis and Repair

Lambda control loop, checking

Before starting to check the control loop be sure that you used the Troubleshooting Chart up to this point.

Test conditions

- Engine oil temperature must be at least 80°C (176°F).
- All electrical consumers such as air conditioner, radio, fan, etc. must be switched off.
- Pinch (close) hose from two-way valve for idle speed boost.

1. Connect the following test equipment:
   - Ignition tester such as VAG 1367
   - Four-gas emission analyzer, such as VAG 1788 connected to CO test connector.

2. Run engine at idle.

3. Increase engine speed to 2500 to 2800 rpm for at least 30 seconds until the Lambda value has stabilized.
   - Read first Lambda value and make note of it.

4. Reduce engine speed to idle and wait at least 30 seconds until the Lambda value has stabilized again.
   - Read second Lambda value and make note of it.

5. Remove crankcase ventilation hose at T-connector (see Fig. 5-11). The Lambda value must increase briefly and then stabilize again at the prior value.
   - If the Lambda value does not change, the control loop does not work properly. Continue in the Troubleshooting Chart at "Control loop not OK". In this case reconnect the disconnected hose.
   - If the Lambda value changes continue as follows.

6. After 60 seconds read the third Lambda value when the Lambda value has stabilized again and make note of it.

7. Reconnect disconnected hose.

8. Reconnect all hoses and lines that were disconnected and secure them properly.

9. Check whether running additional components such as A/C or automatic transmission affects engine performance.

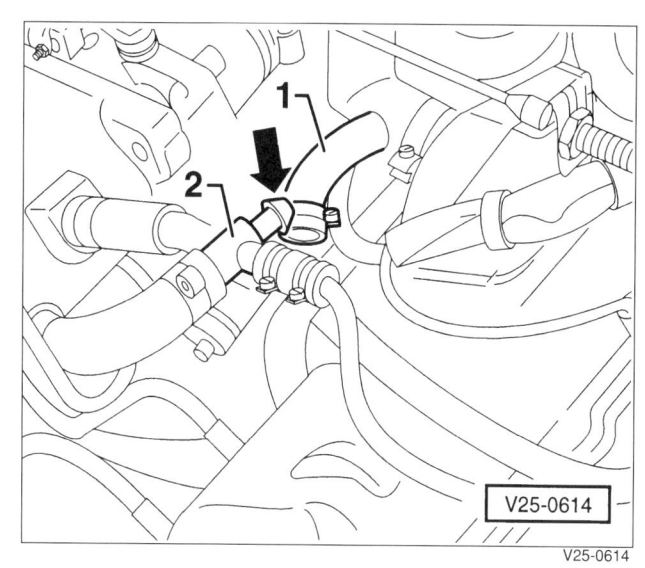

Fig. 5-11. Vacuum hose (**1**) to be disconnected from T-connector (**2**).

10. Evaluate noted Lambda values.
    - If one or more of the noted Lambda values are out of the range 0.97 to 1.03 or if the Lambda value does not stabilize, the control loop does not function properly. Continue in the Troubleshooting Chart at "Control loop not OK".
    - If the noted Lambda values are within 0.97 to 1.03 and if the Lambda value stabilizes, the control loop is OK. Continue in the Troubleshooting Chart at "Control loop OK".

NOTE

To see the relationship between the Lambda value and oxygen sensor output voltage, see Fig. 2-13 in Chapter 2.

ENGINE CODE UM/JN (1.8 liter/4 cylinder, CIS-E)

Control loop, repairing

Equipment needed:

- Ignition tester such as VAG 1367
- CO tester such as VAG 1363 or SUN EPA 75
- Multimeter Fluke 83, US 1119
- Adapter VAG 1315 A/1

1. Perform basic engine adjustments as per Volkswagen Repair Manual (Repair Groups 25 and 28).
   - Check engine idle speed, adjust if necessary.
   - Check CO content, adjust if necessary.

2. If the control loop is still not OK, continue with the troubleshooting.

Checking and Repairing Emission Influencing Components

Check the following components in the table. Perform individual checking steps as per applicable Volkswagen Repair Manual.

NOTE
- *Components that were already recognized as being faulty and were repaired do not have to be rechecked or replaced.*

- *Should you however have found a fault and repaired it, re-check the control loop once again. If the control loop checks out OK, continue in the Troubleshooting Chart with "Control loop OK."*

Additional Components to Test	Volkswagen Repair Manual (Repair manual title and applicable repair group)
Oxygen Sensor and Lambda regulation	**Fuel Injection and Ignition Repair Group 25**
Fuel Injection System	
Air Sensor Plate	
Volume Air Flow Position Sensor	
Thermal Time Switch	
Fuel Pressure Regulator	
Fuel Injectors	
Basic adjustment of throttle valve	
Ignition related faults according to Troubleshooting Chart	**Fuel Injection and Ignition Repair Group 28**

Engine Code KX
(2.2 liter/5 cylinder, CIS-E)

Engines of vehicles listed below have the engine code **KX** and **CIS with Lambda control** and **CIS-E** engine management system.

Engine displacement: 2.2 liter/5 cylinders. The engine is equipped with the emission influencing components listed in **Table o**.

Model	Model Year
Quantum Sedan/Wagon	1987, 1988
Quantum	1983, 1984, 1985, 1986

Table o. Emission Influencing Components: Engine Code KX

Fuel System	CIS (regulated) and CIS-E[1]
Ignition System	Ignition distributor, Vacuum Diaphragm, Engine Control Module (ECM)[2]
Sensors	Volume Air Flow Sensor
	Engine Coolant Temperature Sensor (NTC2)
	Cooling fan switch [3]
	Oxygen Sensor, Heated Oxygen Sensor[4]
	Closed Throttle Position Switch [5]
	Wide Open Throttle Position Switch
Control Devices	EVAP canister purge (vacuum controlled)
	Crankcase ventilation (vacuum controlled)
	Three-Way Catalytic Converter
	Idle speed control via Idle Air Control Valve[6]
Other Components	Solenoid Starting Valve
	Frequency Valve[7]
	Warm-up unit[7]
	Electric Hydraulic Regulator[8]

[1] CIS as well as CIS-E was installed in engines with engine code KX. CIS-E was used as of model year 1985.
[2] Model year 1983: Idle speed controlled by ECM-timing advance controlled by vacuum diaphragm.
[3] Model year 1983 and 1984: Cooling fan switch with CIS (with Lambda regulation).
[4] As of model year 1988: Heated oxygen sensor.
[5] Not model year 1983.
[6] Model year 1983: used when air conditioning is switched on.
[7] Model year 1983 and 1984 with CIS (with Lambda regulation).
[8] As of model year 1985 with change to CIS-E.

Lambda control loop, checking (Vehicles with CIS-E only)

Before starting to check the control loop be sure that you used the Troubleshooting Chart up to this point.

Test conditions

- Engine oil temperature must be at least 80°C (176°F).
- All electrical consumers such as air conditioner, radio, fan, etc. must be switched off.

1. Connect the following test equipment:
   - Multimeter Fluke 83 or US 1119
   - Dual-ended adapter VAG 1490
2. Turn ignition on. Do not start engine.
3. Connect VAG 1490 with I—I (15k ohm) side toward connector for Engine Coolant Temperature Sensor (NTC).
4. Connect multimeter (set to mA) to differential pressure regulator (use VW 1315A/1 test harness).
   - The value displayed on the multimeter must coincide with the value in the table.

Meters above sea level	Specified value
0 m	10 +/- 2 mA
500 m	8.9 +/- 2 mA
1000 m	7.7 +/- 2 mA
1500 m	6.5 +/- 2 mA
2000 m	5.3 +/- 2 mA

5. Disconnect connector for oxygen sensor.
6. Connect green oxygen sensor connector to ground for 20 seconds.
   - The differential pressure regulator current must increase within 20 seconds by about 10 mA (Lambda regulation is OK).
   - If the current does not change, the control loop does not work properly. Continue in the Troubleshooting Chart at "Control loop not OK".

Control loop, repairing

Equipment needed:

- Ignition tester such as VAG 1367
- Adapter US 1112
- CO tester such as VAG 1363 or SUN EPA 75
- Multimeter Fluke 83, US 1119
- Adapter VAG 1315 A/1

1. Perform basic engine adjustment as per Volkswagen Repair Manual (Repair Groups 25 and 28).
   - Check engine idle speed, adjust if necessary.
   - Check CO content, adjust if necessary.
2. If the control loop is still not OK, continue with the troubleshooting.

Emission Diagnosis and Repair

Checking and Repairing Emission Influencing Components

Check the following components in the table. Perform individual checking steps as per applicable Volkswagen Repair Manual.

NOTE
- *Components that were already recognized as being faulty and were repaired do not have to be rechecked or replaced.*
- *Should you however have found a fault and repaired it, re-check the control loop once again. If the control loop checks out OK, continue in the Troubleshooting Chart with "Control loop OK".*

Additional Components to Test	Volkswagen Repair Manual (Repair manual title and applicable repair group)
Oxygen Sensor and Lambda regulation	**Fuel Injection and Ignition Repair Group 25**
Fuel Injection System	
Height Barometer	
Air Sensor Plate	
Volume Air Flow Position Sensor	
Thermal Time Switch	
Fuel Pressure Regulator	
Basic adjustment of throttle valve	
Fuel Injectors	
Ignition related faults according to Troubleshooting Chart	**Fuel Injection and Ignition Repair Group 28**
Camshaft Position Sensor	

ENGINE CODE KX (2.2 liter/5 cylinder, CIS-E)

ENGINE CODE MV
(2.1 liter/4 cylinder, Digifant)

Engines of vehicles listed below have the engine code **MV** and **Digifant** engine management system.

Engine displacement: 2.1 liter/4 cylinders. The engine is equipped with the emission influencing components listed in **Table p.**

Model	Model Year
Vanagon	1986, 1987, 1988, 1989, 1990, 1991

Table p. Emission Influencing Components: Engine Code MV

Fuel System	Digifant MFI
Ignition System	Ignition distributor, Electronic Control Module (ECM)
Sensors	Volume Air Flow Sensor
	Intake Air Temperature Sensor (NTC1)
	Engine Coolant Temperature Sensor (NTC2)
	Heated Oxygen Sensor
	Closed Throttle Position Switch
	Wide Open Throttle Switch
Control Devices	EVAP canister purge (vacuum controlled)
	Crankcase ventilation (vacuum controlled)
	Three-Way Catalytic Converter
	Automatic Idle speed control
Other Components	Inlet restrictor

Emission Diagnosis and Repair

Lambda control loop, checking

Before checking the control loop, be sure that you used the Troubleshooting Chart up to this point. Read Diagnostic Trouble Code (DTC) memory and Output Diagnostic Test Mode (DTM) using the VAG 1551 fault read out unit. If faults are found, correct them first and then clear the fault memory. See applicable Volkswagen Repair Manual.

Test conditions

- Engine oil temperature must be at least 80°C (176°F).
- All electrical consumers such as air conditioner, radio, fan, etc. must be switched off.

1. Connect the following test equipment:
   - Ignition tester such as VAG 1367
   - Four-gas emission analyzer, such as VAG 1788 connected to CO test connector.

2. Run engine at operating temperature.

3. Increase engine speed to 2500 to 2800 rpm for at least 30 seconds until the Lambda value has stabilized.
   - Read first Lambda value and make note of it.

4. Reduce engine speed to idle and wait at least 30 seconds until the Lambda value has stabilized again.
   - Read second Lambda value and make note of it.

5. Remove vacuum hose from fuel pressure regulator. Close off hose connector for 30 seconds and open it again. See Fig. 5-12. The Lambda value must increase briefly and then stabilize again at the prior value.
   - If the Lambda value does not change, the control loop does not work properly. Continue in the Troubleshooting Chart at "Control loop not OK". In this case reconnect disconnected hose.
   - If the Lambda value briefly increases and then decreases, the control loop is OK. Continue as follows.

6. After 60 seconds read the third Lambda value and make note of it.

7. Reconnect the vacuum hose and all hoses and lines that were disconnected and secure them properly.

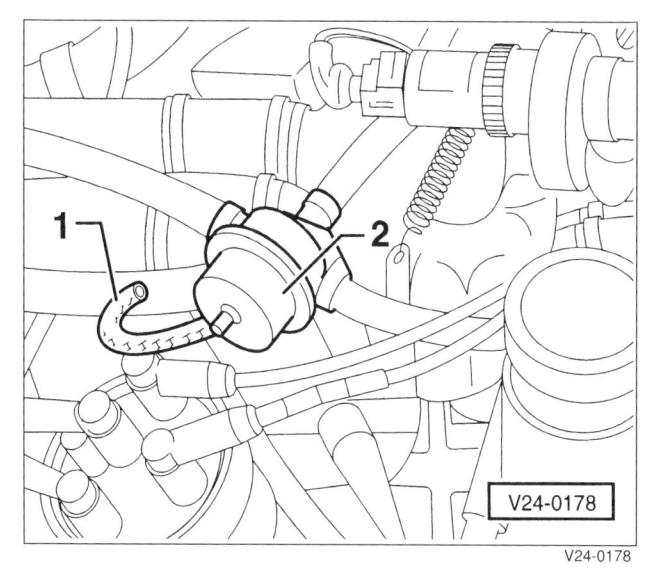

Fig. 5-12. Vacuum hose (**1**) to be removed from fuel pressure regulator (**2**).

8. Check whether running additional components such as A/C or automatic transmission affects engine performance.

9. Evaluate noted Lambda values.
   - If one or more of the noted Lambda values are out of the range 0.97 to 1.03 or if the Lambda value does not stabilize, the control loop does not function properly. Continue in the Troubleshooting Chart at "Control loop not OK".
   - If the noted Lambda values are within 0.97 to 1.03 and if the Lambda value stabilizes, the control loop is OK. Continue in the Troubleshooting Chart at "Control loop OK".

NOTE

To see the relationship between the Lambda value and oxygen sensor output voltage, see Fig. 2-13 in Chapter 2.

ENGINE CODE MV (2.1 liter/4 cylinder, Digifant)

Control loop, repairing

Equipment needed:

- Ignition tester such as VAG 1367
- Adapter such as VAG 1473
- CO tester such as VAG 1363 or SUN EPA 75
- Multimeter Fluke 83, US 1119
- Adapter VAG 1315 A/2

1. Perform basic engine adjustment as per Volkswagen Repair Manual (Repair Groups 25 and 28).
   - Check engine idle speed, adjust if necessary.
   - Check CO content, adjust if necessary.

2. If the control loop is still not OK, continue with the following steps.

Checking and Repairing Emission Influencing Components

Check the following components. Perform individual checking steps as per applicable Volkswagen Repair Manual.

NOTE
- *Components that were already recognized as being faulty and were repaired do not have to be rechecked or replaced.*
- *Should you however have found a fault and repaired it, re-check the control loop once again. If the control loop checks out OK, continue in the Troubleshooting Chart with "Control loop OK."*

Individual component checks as per Volkswagen Repair Manual, Repair Group 24:

- Voltage supply to coil
- Voltage supply to ECM (J169)
- Fuel injectors (N30...N33)
- Temperature Sensor II (G18)
- Throttle Position Switches (F25 or F60/F81)
- Volume Air Flow Sensor Position Sensor (G19)
- Intake Air Temperature Sensor (G42)
- Wiring to Heated Oxygen Sensor (G39)

If the control loop is still not OK, continue with checking the following components listed in the table:

Additional Components to Test	Volkswagen Repair Manual (Repair manual title and applicable repair group)
Volume Air Flow Sensor	Fuel Injection and Ignition Repair Group 24
Fuel Pressure Regulator	
Oxygen Sensor	
Fuel Injectors	
Closed Throttle Position Switch (F60) and Wide Open Throttle Witch (F81)	
Idle Air Control Valve	
Basic adjustment of throttle valve	
Ignition timing	Fuel Injection and Ignition Repair Group 28

ENGINE CODE MV (2.1 liter/4 cylinder, Digifant)

Engine Code PF
(1.8 liter/4 cylinder, Digifant II)

Engines of vehicles listed below have the engine code **PF** and **Digifant II** engine management system.

Engine displacement: 1.8 liter/4 cylinders. The engine is equipped with the emission influencing components listed in **Table q**.

Model	Model Year
GTI (8-valve)	1991, 1992
Golf	1987, 1988, 1989, 1990, 1991
Jetta	1987, 1988, 1989, 1990, 1991

Table q. Emission Influencing Components: Engine Code PF

Fuel System	Digifant II
Ignition System	Distributor ignition[3], Electronic Control Module (ECM)
Sensors	Volume Air Flow Sensor
	Intake Air Temperature Sensor (NTC1)
	Engine Coolant Temperature Sensor (NTC2)
	Heated Oxygen Sensor
	Closed Throttle Position Switch/Throttle Position Sensor[2]
	Wide Open Throttle Position Switch/Throttle Position Sensor[2]
	Knock Sensor
Control Devices	EVAP canister purge (vacuum controlled)
	Crankcase ventilation (vacuum controlled)
	Three-Way Catalytic Converter
	Automatic Idle speed control
	Diagnostic Trouble Code: Blink Code[1]
Other Components	Inlet restrictor

[1] For California model only, Digifant I.
[2] For California, model year 1991 only, Digifant II.
[3] California, model year 1991, Digifant II uses another style of ignition coil.

Lambda control loop, checking

Before checking the control loop, be sure that you used the Troubleshooting Chart up to this point. Read Diagnostic Trouble Code (DTC) memory and Output Diagnostic Test Mode (DTM) using the VAG 1551 fault read out unit. If faults are found, correct them first and then clear the fault memory. See applicable Volkswagen Repair Manual.

Test conditions

- Engine oil temperature must be at least 80°C (176°F).
- All electrical consumers such as air conditioner, radio, fan, etc. must be switched off.
- Idle Air Control must be OK (with ignition turned on, the Idle Air Control Valve must buzz).

1. Connect the following test equipment:
   - Ignition tester such as VAG 1367
   - Four-gas analyzer, such as VAG 1788 connected to CO test connector.

2. Run engine at idle (engine must be at operating temperature).

3. Increase engine speed to 2500 to 2800 rpm for at least 30 seconds until the Lambda value has stabilized.
   - Read first Lambda value and make note of it.

4. Reduce engine speed to idle and wait at least 30 seconds until the Lambda value has stabilized again.
   - Read second Lambda value and make note of it.

5. Remove hose from fuel pressure regulator at throttle valve housing. See Fig. 5-13. Close hose connector for 30 seconds and open it again. The Lambda value must increase briefly and then stabilize again at the prior value.
   - If the Lambda value does not change, the control loop does not work properly. Continue in the Troubleshooting Chart at "Control loop not OK". In this case reconnect hose again.
   - If the Lambda value briefly increases and then decreases, the control loop is OK. Continue as follows.

6. After 60 seconds read the third Lambda value and make note of it.

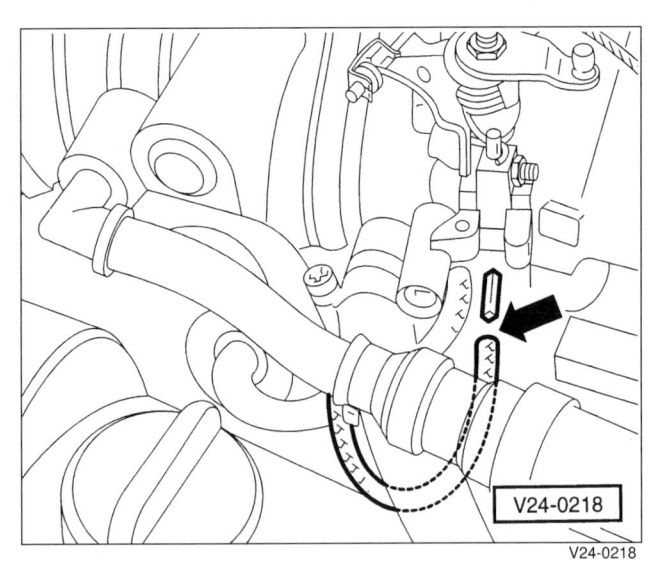

Fig. 5-13. Vacuum hose (arrow) to be disconnected at throttle valve housing.

7. Reconnect hose and all hoses and lines that were disconnected and secure them properly.

8. Check whether running additional components such as A/C or automatic transmission affects engine performance.

9. Evaluate noted Lambda values
   - If one or more of the noted Lambda values are out of the range 0.97 to 1.03 or if the Lambda value does not stabilize, the control loop does not function properly. Continue in the Troubleshooting Chart at "Control loop not OK".
   - If the noted Lambda values are within 0.97 to 1.03 and if the Lambda value stabilizes, the control loop is OK. Continue in the Troubleshooting Chart at "Control loop OK".

ENGINE CODE PF (1.8 liter/4 cylinder, Digifant II)

Emission Diagnosis and Repair

Control loop, repairing

Equipment needed:

- Ignition tester such as VAG 1367
- CO tester such as VAG 1363 or SUN EPA 75
- Special tool 6006-0019

1. Perform basic engine adjustment as per Volkswagen Repair Manual (Repair Group 24).
   - Check engine idle speed, adjust if necessary.
   - Check CO content, adjust if necessary.

2. If the control loop is still not OK, continue with the troubleshooting.

Checking and Repairing Emission Influencing Components

Check the following components in the table. Perform individual checking steps as per applicable Volkswagen Repair Manual.

NOTE
- *Components that were already recognized as being faulty and were repaired do not have to be rechecked or replaced.*
- *Should you however have found a fault and repaired it, re-check the control loop once again. If the control loop checks out OK, continue in the Troubleshooting Chart with "Control loop OK."*

Additional Components to Test	Volkswagen Repair Manual (Repair manual title and applicable repair group)
Air Sensor Plate	**Fuel Injection and Ignition Repair Group 24**
Volume Air Flow Position Sensor	
Fuel Pressure Regulator	
Oxygen Sensor and Lambda regulation	
Crankcase Ventilation Valve	
Fuel Injectors	
Closed Throttle and Wide Open Throttle Position Switches	
Throttle Position Switch	
Ignition system related problems	**Fuel Injection and Ignition Repair Group 28**

ENGINE CODE PF (1.8 liter/4 cylinder, Digifant II)

Engine Code PG
(1.8 liter/4 cylinder, Digifant)

Engines of vehicles listed below have the engine code **PG** and **Digifant** engine management system. Engine displacement: 1.8 liter/4 cylinders. The engine is equipped with the emission influencing components listed in **Table r**.

Model	Model Year
Corrado/G60	1989, 1990, 1991, 1992

Table r. Emission Influencing Components: Engine Code PG

Fuel System	Digifant MFI
Ignition System	Ignition distributor, Electronic Control Module (ECM)
Sensors	Intake Air Temperature Sensor (NTC1)
	Engine Coolant Temperature Sensor (NTC2)
	Heated Oxygen Sensor
	Closed Throttle Position Switch[1], Throttle Position Sensor [2,3]
	Wide Open Throttle Switch[1,3]
	Knock Sensor
Control Devices	Vacuum controlled EVAP canister purge
	Vacuum controlled crankcase ventilation
	Three-Way Catalytic Converter
	Automatic Idle speed control
	Diagnostic Trouble Code: MIL lamp
Other Components	Inlet restrictor

[1] For USA, except California
[2] For California only
[3] For California models, MY 1991 with Closed Throttle Position Sensor.

Emission Diagnosis and Repair

Lambda control loop, checking

Before checking the control loop, be sure that you used the Troubleshooting Chart up to this point. Read Diagnostic Trouble Code (DTC) memory and Output Diagnostic Test Mode (DTM) using the VAG 1551 fault read out unit. If faults are found, correct them first and then clear the fault memory. See applicable Volkswagen Repair Manual.

Test conditions

- Engine oil temperature must be at least 80°C (176°F).
- All electrical consumers such as air conditioner, radio, fan, etc. must be switched off.
- Idle Air Control must be OK (with ignition turned on, the Idle Air Control Valve must buzz).

1. Connect the following test equipment:
   - Volkswagen fault read out unit, VAG 1551
   - Ignition tester such as VAG 1367
   - Four-gas analyzer, such as VAG 1788 to CO test connector.

2. Run engine at idle (engine must be at operating temperature).

3. Increase engine speed to 2500 to 2800 rpm for at least 30 seconds until the Lambda value has stabilized.
   - Read first Lambda value and make note of it.

4. Reduce engine speed to idle and wait at least 30 seconds until the Lambda value has stabilized again.
   - Read second Lambda value and make note of it.

5. Remove small vacuum hose from intake manifold. See Fig. 5-14. Immediately close off hose connector. The Lambda value must decrease briefly and then increase again after 60 seconds.

6. If the Lambda value does not change or did not stabilize again at the previous value, the control loop does not work properly. Continue in the Troubleshooting Chart at "Control loop not OK". In this case reconnect hose again.
   - If the Lambda value briefly decreases and then increases, the control loop is OK. Continue as follows.

7. Read the third Lambda value and make note of it.

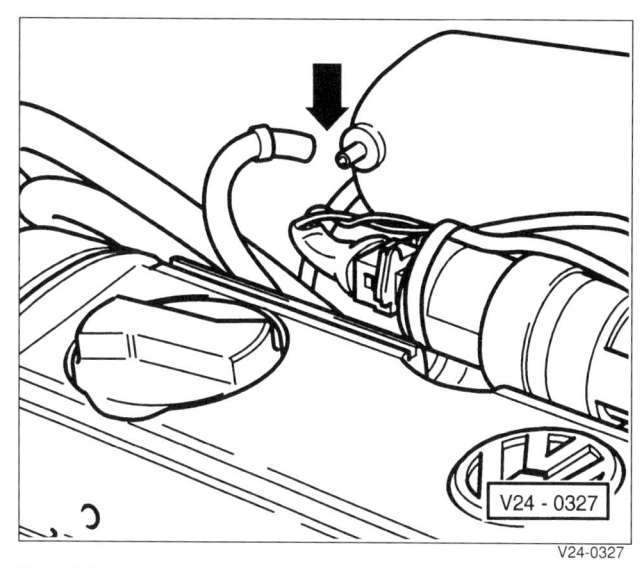

Fig. 5-14. Vacuum hose (arrow) at intake manifold to be disconnected.

8. Reconnect hose and all hoses and lines that were disconnected and secure them properly.

9. Check whether running additional components such as A/C or automatic transmission affects engine performance.

10. Evaluate noted Lambda values.
    - If one or more of the noted Lambda values are out of the range 0.97 to 1.03 or if the Lambda value does not stabilize, the control loop does not function properly. Continue in the Troubleshooting Chart at "Control loop not OK".
    - If the noted Lambda values are within 0.97 to 1.03 and if the Lambda value stabilizes, the control loop is OK. Continue in the Troubleshooting Chart at "Control loop OK".

NOTE

To see the relationship between the Lambda value and oxygen sensor output voltage, see Fig. 2-13 in Chapter 2.

ENGINE CODE PG (1.8 liter/4 cylinder, Digifant)

Control loop, repairing
(Vehicles for USA as of MY 1989 and Calif. vehicles MY 1989–1990 w/Digifant II)

Equipment needed:

- Ignition tester such as VAG 1367
- Adapter such as VAG 1363/3
- CO tester such as VAG 1363 or SUN 105
- Multimeter Fluke 83, US 1119 with adapter 1315A/2

1. Perform basic engine adjustment as per Volkswagen Repair Manual, Fuel Injection and Ignition (Repair Groups 24 and 28).

2. Check engine idle speed, adjust if necessary.
   - Check CO content, adjust if necessary.
   - Check ignition timing.

3. If the control loop is still not OK, continue with the troubleshooting described below.

Control loop check
(Vehicles for California MY 1991 w/Digifant I)

Equipment needed:

- Ignition tester such as VAG 1367
- Trigger pliers such as VAG 1367/8
- Adapter such as VAG 1363/3
- CO tester such as VAG 1363 or SUN 105
- Multimeter Fluke 83, US 1119 with adapter 1315A/2

1. Perform basic engine adjustment as per Volkswagen Repair Manual (Repair Groups 24 and 28).

> *Caution—*
> *Ignition timing, CO content and idle speed must always be checked and adjusted together.*

   - Check engine idle speed, adjust if necessary.
   - Check CO content, adjust if necessary.
   - Check ignition timing.

2. If the control loop is still not OK, continue with the troubleshooting describe below.

ENGINE CODE PG (1.8 liter/4 cylinder, Digifant)

Emission Diagnosis and Repair

Checking and Repairing Emission Influencing Components

Check the following components in the table. Perform individual checking steps as per applicable Volkswagen Repair Manual.

NOTE
- *Components that were already recognized as being faulty and were repaired do not have to be rechecked or replaced.*
- *Should you however have found a fault and repaired it, re-check the control loop once again. If the control loop checks out OK, continue in the Troubleshooting Chart with "Control loop OK."*
- *Components that apply to both Digifant I and Digifant II are not specifically identified. Components that are specific to a particular engine management system are identified in brackets.*

Additional Components to Test	Volkswagen Repair Manual (Repair manual title and applicable repair group)
Oxygen Sensor and Lambda regulation	**Fuel Injection and Ignition Repair Group 24**
Volume Air Flow Position Sensor	
Fuel Pressure Regulator	
Fuel Injectors	
Throttle Position Switch	
Closed Throttle and Wide Open Throttle Position Switches (Digifant II)	
CO-potentiometer (Digifant I)	
Idle Air Control Valve	
Basic adjustment of throttle valve	
Intake Air Temperature Sensor (Digifant I)	
Throttle Position Sensor (Digifant I)	
Knock Sensor (Digifant II)	**Fuel Injection and Ignition Repair Group 28**
Ignition timing	

ENGINE CODE PG (1.8 liter/4 cylinder, Digifant)

Engine Code PL
(1.8 liter/4 cylinders, CIS-E)

Engines of vehicles listed below have the engine code **PL** and **CIS-E** engine management system.

Engine displacement: 1.8 liter/4 cylinders, 16V. The engine is equipped with the emission influencing components listed in **Table s**.

Model	Model Year
Scirocco	1986, 1987, 1988, 1989
Jetta GLI	1987, 1988, 1989
GTI	1987, 1988, 1989

Table s. Emission Influencing Components: Engine Code PL

Fuel system	CIS-E
Ignition System	Ignition distributor, Electronic Control Module (ECM)
Sensors	Volume Air Flow Sensor
	Engine Coolant Temperature Sensor (NTC2)
	Heated Oxygen Sensor
	Closed Throttle Position Switch
	Wide Open Throttle Position Switch
	Knock Sensor
Control Devices	Vacuum controlled EVAP canister purge
	EVAP canister: located in right front wheel housing
	Vacuum controlled crankcase ventilation
	Three-Way Catalytic Converter
	Idle speed control: Idle Air Control Valve
	Diagnostic Trouble Code: Ignition System
Other Components	Solenoid Starting Valve
	Electrical-Hydraulic Regulator

Emission Diagnosis and Repair

Lambda control loop, checking

Before starting the control loop be sure that you used the Troubleshooting Chart up to this point. Read DTC memory and Output Diagnostic Test Mode (DTM) for California vehicles. If you found faults, correct them first and then erase the memory, Repair Manual "On Board Diagnostics Troubleshooting, Group 01".

Test conditions

- Engine oil temperature must be at least 80°C (176°F).
- All electrical consumers such as air conditioner, radio, fan, etc. must be switched off.

1. Connect the following test equipment:
   - Ignition tester such as VAG 1367
   - Four-gas emission analyzer, such as VAG 1788 connected to CO test connector.

2. Run engine at idle (engine must be at operating temperature).

3. Increase engine speed to 2500 to 2800 rpm for at least 30 seconds until the Lambda value has stabilized.
   - Read first Lambda value and make note of it.

4. Reduce engine speed to idle and wait at least 30 seconds until the Lambda value has stabilized again.
   - Read second Lambda value and make note of it.

5. Remove hose for crankcase ventilation with connector. See Fig. 5-15.

6. Open hose connector. The Lambda value must increase briefly and then stabilize again at the prior value.
   - If the Lambda value does not change the control loop does not work properly. Continue in the Troubleshooting Chart at "Control loop not OK". In this case reconnect crankcase ventilation hose.
   - If the Lambda value briefly decreases and then increases, the control loop is OK. Continue as follows.

7. After 60 seconds read the third Lambda value and make note of it.

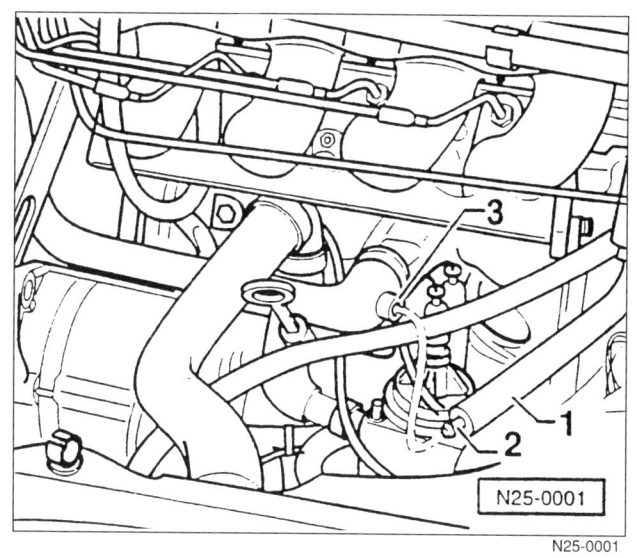

Fig. 5-15. Disconnect crankcase ventilation hose (**1**) at hose connection (**3**).

8. Reconnect crankcase ventilation hose and all hoses and lines that were disconnected and secure them properly.

9. Check whether running additional components such as A/C or automatic transmission affects engine performance.

10. Evaluate noted Lambda values
    - If one or more of the noted Lambda values are out of the range 0.97 to 1.03 or if the Lambda value does not stabilize, the control loop does not function properly. Continue in the Troubleshooting Chart at "Control loop not OK".
    - If the noted Lambda values are within 0.97 to 1.03 and if the Lambda value stabilizes, the control loop is OK. Continue in the Troubleshooting Chart at "Control loop OK".

NOTE

To see the relationship between the Lambda value and oxygen sensor output voltage, see Fig. 2-13 in Chapter 2.

ENGINE CODE PL (1.8 liter/4 cylinders, CIS-E)

Control loop, repairing

Equipment needed:

- Ignition tester such as VAG 1367
- Trigger pliers such as VAG 1367/8
- Adapter such as VAG 1363/3
- CO tester such as VAG 1363 or SUN EPA 75

1. Perform basic engine adjustment as per Volkswagen Repair Manual (Repair Groups 25 and 28).
   - Check engine idle speed, adjust if necessary.
   - Check CO content, adjust if necessary.
   - Check ignition timing.

2. If the control loop is still not OK, continue with the troubleshooting.

Checking and Repairing Emission Influencing Components

Check the following components in the table. Perform individual checking steps as per applicable Volkswagen Repair Manual.

NOTE
- *Components that were already recognized as being faulty and were repaired do not have to be rechecked or replaced.*
- *Should you however have found a fault and repaired it, re-check the control loop once again. If the control loop checks out OK, continue in the Troubleshooting Chart with "Control loop OK."*

Component to Check	Repair Manual and applicable Repair Group
Oxygen Sensor and Lambda regulation	**Fuel Injection and Ignition Repair Group 25**
Wide Open Throttle Position Switch	
Closed Throttle Position Switch	
Air Sensor Plate	
Cold Start Injector	
Thermal Time Switch	
Volume Air Flow Position Sensor	
Fuel Pressure Regulator	
Differential Pressure Regulator	
Fuel Injectors	
Intake air pre-heating	
Knock Sensor (Digifant II)	**Fuel Injection and Ignition Repair Group 28**
Ignition timing	

ENGINE CODE PL (1.8 liter/4 cylinders, CIS-E)

Emission Diagnosis and Repair

ENGINE CODE RD (1.8 liter/4 cylinders, CIS-E)

Engines of vehicles listed below have the engine code **RD** and **CIS-E** engine management system. Engine displacement: 1.8 liter/4 cylinders. The engine is equipped with the emission influencing components listed in **Table t**.

Model	Model Year
Golf GT	1986, 1987, 1988
Golf GTI (8-valve)	1986, 1987, 1988
Jetta GLI (8-valve)	1986, 1987, 1988

Table t. Emission Influencing Components: Engine Code RD

Fuel System	CIS-E
Ignition System	Ignition distributor, Electronic Control Module (ECM)
Sensors	Volume Air Flow Sensor
	Engine Coolant Temperature Sensor (NTC2)
	Heated Oxygen Sensor
	Closed Throttle Position Switch
	Wide Open Throttle Position Switch
	Knock Sensor
Control Devices	EVAP canister purge (vacuum controlled)
	EVAP canister: located in right front wheel housing
	Crankcase ventilation (vacuum controlled)
	Three-Way Catalytic Converter
	Idle speed control: Idle Air Control Valve
	Diagnostic Trouble Code: Ignition System
Other Components	Solenoid Starting Valve
	Electric hydraulic regulator (Differential Pressure Regulator)

Lambda control loop, checking

Before starting the control loop be sure that you used the Troubleshooting Chart up to this point.

Test conditions

- Engine oil temperature must be at least 80°C (176°F).
- All electrical consumers such as air conditioner, radio, fan, etc. must be switched off.

1. Connect the following test equipment:
   - Ignition tester such as VAG 1367
   - Four-gas emission analyzer, such as VAG 1788 connected to CO test connector.

2. Increase engine speed to 2500 to 2800 rpm for at least 30 seconds until the Lambda value has stabilized.
   - Read first Lambda value and make note of it.

3. Reduce engine speed to idle and wait at least 30 seconds until the Lambda value has stabilized again.
   - Read second Lambda value and make note of it.

4. Remove crankcase ventilation vacuum hose from connector on intake manifold. Close off hose connector for 30 seconds and open it again. (see Fig. 5-16). The Lambda value must increase briefly and then stabilize again at the prior value.
   - If the Lambda value does not change, the control loop does not work properly. Continue in the Troubleshooting Chart at "Control loop not OK". In this case reconnect disconnected hose.
   - If the Lambda value briefly increases and then decreases then stabilize again at the prior value, the control loop is OK. Continue as follows.

5. After 60 seconds, read the third Lambda value and make note of it.

6. Reconnect vacuum hose and all hoses and lines that were disconnected and secure them properly.

7. Check whether running additional components such as A/C or automatic transmission affects engine performance.

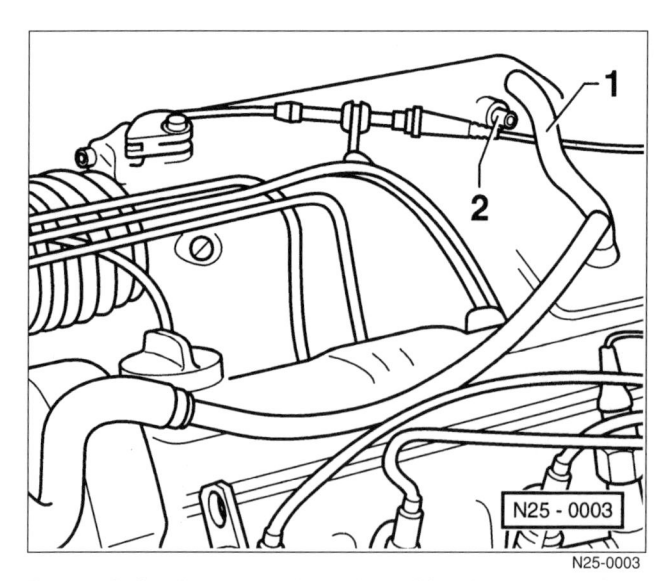

Fig. 5-16. Crankcase ventilation hose (**1**) to be removed from intake manifold (**2**).

8. Evaluate noted Lambda values
   - If one or more of the noted Lambda values are out of the range 0.97 to 1.03 or if the Lambda value does not stabilize, the control loop does not function properly. Continue in the Troubleshooting Chart at "Control loop not OK".
   - If the noted Lambda values are within 0.97 to 1.03 and if the Lambda value stabilizes, the control loop is OK. Continue in the Troubleshooting Chart at "Control loop OK".

NOTE

To see the relationship between the Lambda value and oxygen sensor output voltage, see Fig. 2-13 in Chapter 2.

ENGINE CODE RD (1.8 liter/4 cylinders, CIS-E)

Emission Diagnosis and Repair

Control loop, repairing

Equipment needed:

- Ignition tester such as VAG 1367
- Trigger pliers such as US 1112
- Adapter such as VAG 1363/3
- CO tester such as VAG 1363 or SUN 105
- Multimeter Fluke 83, US 1119
- Adapter VAG 1315 A/1

1. Perform basic engine adjustment as per Volkswagen Repair Manual (Repair Groups 25 and 28).
   - Check engine idle speed, adjust if necessary.
   - Check CO content, adjust if necessary.
2. If the control loop is still not OK, continue with the troubleshooting.

Checking and Repairing Emission Influencing Components

Check the following components listed below and the components given in the table on the following page. Perform individual checking steps as per applicable Volkswagen Repair Manual.

NOTE

- *Components that were already recognized as being faulty and were repaired do not have to be rechecked or replaced.*

- *Should you however have found a fault and repaired it, re-check the control loop once again. If the control loop checks out OK, continue in the Troubleshooting Chart with "Control loop OK."*

- Idle Air Control Valve (N71), Repair Group 25
- Volume Air Flow Sensor Position Sensor (G19), Repair Group 25
- Engine Coolant Temperature Sensor (N10), Repair Group 25
- Power supply for Ignition Control Module (J154), Repair Group 28
- Wide Open Throttle Switch (F81), Repair Groups 25 and 28
- Closed Throttle Position Switch (F60), Repair Groups 25 and 28
- Wiring to Engine Control Module (ECM) (J21), pin 25, Repair Group 28
- Wiring for On Board Diagnostics, Repair Group 28

Additional Components to Test	Volkswagen Repair Manual (Repair manual title and applicable repair group)
Oxygen Sensor and Lambda regulation	**Fuel Injection and Ignition Repair Group 25**
Fuel Injectors	
Closed Throttle Position Switch (F60) and Wide Open Throttle Witch (F81)	
Air Sensor Plate	
Volume Air Flow Position Sensor	
Cold Start Injector (N17)	
Cold Start Injector Thermal Switch (F26)	
Fuel Pressure Regulator	
Differential Pressure Regulator	
Ignition Timing	**Fuel Injection and Ignition Repair Group 28**

ENGINE CODE RD (1.8 liter/4 cylinders, CIS-E)

ENGINE CODE RV
(1.8 liter/4 cylinders, Digifant)

Engines of vehicles listed below have the engine code **RV** and **Digifant** engine management system.

Engine displacement: 1.8 liter/4 cylinders. The engine is equipped with the emission influencing components listed in **Table u**.

Model	Model Year
Golf	1987, 1988, 1989, 1990, 1991, 1992
Golf GTI	1991, 1992
Jetta	1987, 1988, 1989, 1990, 1991, 1992

Table u. Emission Influencing Components: Engine Code RV

Fuel System	Digifant MFI
Ignition System	Ignition distributor, Electronic Control Module (ECM)
Sensors	Volume Air Flow Sensor
	Intake Air Temperature Sensor (NTC1)
	Engine Coolant Temperature Sensor (NTC2)
	Heated Oxygen Sensor
	Closed Throttle Position Switch[1], Throttle Position Sensor[2]
	Wide Open Throttle Switch[2]
	Knock Sensor
Control Devices	EVAP canister purge (vacuum controlled)
	Crankcase ventilation (vacuum controlled
	Three-Way Catalytic Converter
	Automatic Idle speed control
	Exhaust gas recirculation (EGR) Valve, EGR Temperature Sensor[1]
	Diagnostic Trouble Code: Rapid Data Transfer[4], MIL lamp[3]
Other Components	Inlet restrictor

[1] For USA, except California
[2] For California only, MY 1991
[3] For California only, MY 1991
[4] On Board Diagnostics (OBD) for Jetta, model years 1988–1990

Emission Diagnosis and Repair 113

Lambda control loop, checking

Before checking the control loop, be sure that you used the Troubleshooting Chart up to this point. Read Diagnostic Trouble Code (DTC) memory and Output Diagnostic Test Mode (DTM) using the VAG 1551 fault read out unit. If faults are found, correct them first and then clear the fault memory. See applicable Volkswagen Repair Manual.

Test conditions

- Engine oil temperature must be at least 80°C (176°F).
- All electrical consumers such as air conditioner, radio, fan, etc. must be switched off.

1. Connect the following test equipment:
   - Four-gas emission analyzer, such as VAG 1788 or CO tester such as VAG 1363 with adapter 1363/3 connected to CO test connector.

2. Run engine at idle (engine at operating temperature).
   - Read first Lambda value and make note of it.

3. Remove hose from fuel pressure regulator at throttle housing and close small tubing. See Fig. 5-17. The CO content must briefly increase and then decrease and stabilize at the noted value.
   - If the Lambda value does not change, the control loop does not work properly. Continue in the Troubleshooting Chart at "Control loop not OK". In this case reconnect disconnected hose.
   - If the CO content briefly increases and then decreases and stabilizes again at the prior value, the control loop is OK. Continue as follows.

4. Reconnect all hoses and lines that were disconnected and secure them properly.

5. Check whether running additional components such as A/C or automatic transmission affects engine performance.

NOTE

If the Lambda value does not change, the control loop does not work properly. Continue with the troubleshooting described below.

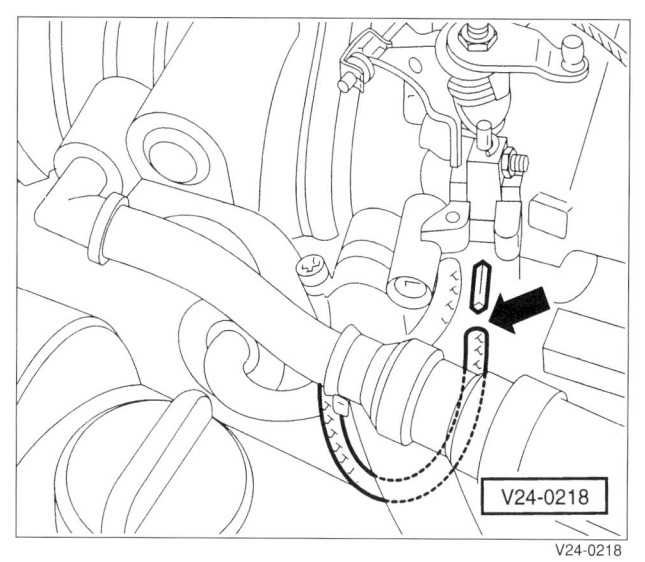

Fig. 5-17. Vacuum hose (arrow) to be removed from throttle valve housing.

6. Remove connector of oxygen sensor and alternately connect wiring to ECM to ground and plus (+) voltage (use 1.5 volt dry-cell for plus (+) voltage). See Fig. 5-18.
   - CO content must increase or decrease.
   - If CO content changes, replace oxygen sensor.
   - If CO content does not change, check wiring to ECM and repair if necessary.

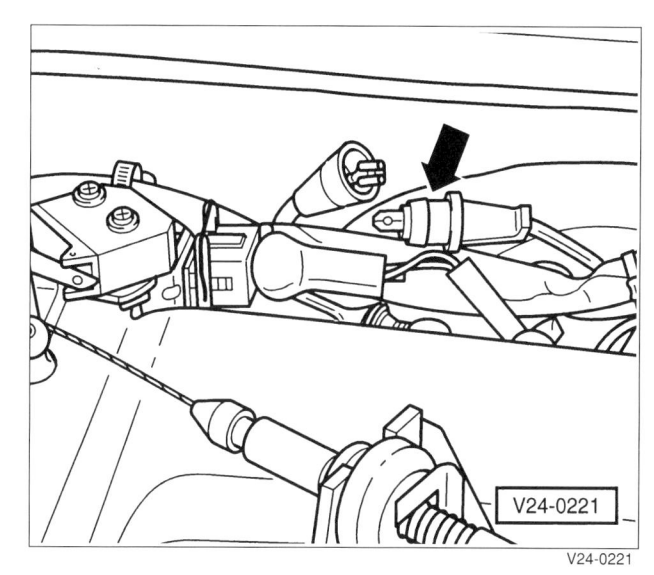

Fig. 5-18. Oxygen sensor connector (arrow) to be disconnected.

7. Reconnect all hoses and lines that were disconnected and secure them properly.

ENGINE CODE RV (1.8 liter/4 cylinders, Digifant)

8. Check whether running additional components such as A/C or automatic transmission affects engine performance.

9. Evaluate noted Lambda values
   - If one or more of the noted Lambda values are out of range or if the Lambda value does not stabilize, the control loop does not function properly. Continue in the Troubleshooting Chart at "Control loop not OK".
   - If the noted Lambda values are within range and if the Lambda value stabilizes, the control loop is OK. Continue in the Troubleshooting Chart at "Control loop OK".

NOTE
To see the relationship between the Lambda value and oxygen sensor output voltage, see Fig. 2-13 in Chapter 2.

Control loop, repairing (Vehicles for USA vehicles as of MY1991 Digifant II)

Test conditions

- Engine oil temperature must be at least 80°C (176°F).
- All electrical consumers such as air conditioner, radio, fan, etc. must be switched off.

Equipment needed:

- Ignition tester such as VAG 1367
- CO tester such as VAG 1363 or SUN 105
- Special tool 6006-0019

1. Perform basic engine adjustment as per Volkswagen Repair Manual (Repair Groups 25 and 28, for **Engine Code PF**).
   - Check engine idle speed, adjust if necessary.
   - Check CO content, adjust if necessary.
   - Check ignition timing.

2. If the control loop is still not OK, continue with the troubleshooting under **Checking and Repairing Emission Influencing Components**.

Control loop repairing (Vehicles for California as of m. y. 1991 Digifant I)

Test conditions

- Engine oil temperature must be at least 80°C (176°F).
- All electrical consumers such as air conditioner, radio, fan, etc. must be switched off.

Equipment needed:

- Ignition tester such as VAG 1367
- Trigger pliers such as VAG 1367/8
- Adapter such as VAG 1363 or SUN 105
- CO tester such as VAG 1363 or SUN 105

1. Perform basic engine adjustment as per Volkswagen Repair Manual (Repair Groups 24 and 28).
   - Check engine idle speed, adjust if necessary.
   - Check CO content, adjust if necessary.
   - Check ignition timing.

2. If the control loop is still not OK, continue with the troubleshooting under **Checking and Repairing Emission Influencing Components**.

ENGINE CODE RV (1.8 liter/4 cylinders, Digifant)

Emission Diagnosis and Repair

Checking and Repairing Emission Influencing Components

Check the following components in the table. Perform individual checking steps as per applicable Volkswagen Repair Manual.

NOTE

- *Components that were already recognized as being faulty and were repaired do not have to be rechecked or replaced.*

- *Should you however have found a fault and repaired it, re-check the control loop once again. If the control loop checks out OK, continue in the troubleshooting chart with "Control loop OK."*

- *Components that apply to Digifant I and Digifant II are not specifically identified. Components that apply to Digifant I and Digifant II are identified in brackets.*

Additional Components to Test	Volkswagen Repair Manual (Repair manual title and applicable repair group)
Volume Air Flow Sensor	**Fuel Injection and Ignition Repair Group 24**
Fuel Pressure Regulator	
Oxygen Sensor and Lambda regulation	
Volume Air Flow Position Sensor (Digifant I)	
Closed Throttle and Wide Open Throttle Position Switches (Digifant II)	
Engine Coolant Temperature Sensor (Digifant I)	
Throttle Position Switch (Digifant I)	
Fuel Injectors	
Throttle Position Sensor	
Idle Air Control Valve	
Basic adjustment of throttle valve	
CO potentiometer (Digifant I)	
Intake Air Pressure Sensor (Digifant I)	
Camshaft Position Sensor	
Knock Sensor (Digifant II)	**Fuel Injection and Ignition Repair Group 28**
Ignition timing	

ENGINE CODE RV (1.8 liter/4 cylinders, Digifant)

Appendix A

Engine Application Chart

> *NOTE—*
> *It is recommended that the engine code given in the table below be verified by cross checking it with the actual code on the engine itself. Some Volkswagen models may have been manufacturered with more than one engine code during a given model year. See Fig. 5-1 found at the beginning of Chapter 5.*

Model /Years	Engine Code	Displacement	Engine Management/ Fuel Delivery System	Page #
Cabrio				
1994–1997	ABA	2.0 Liter / 4-cylinder	Motronic	58
Cabriolet (see also Rabbit Convertible)				
1985–1989	JH	1.8 Liter / 4-cylinder	CIS with Lambda regulation	87
1990–1993	2H	1.8 Liter / 4-cylinder	Digifant II, Digifant I	68
Corrado				
G60				
1989–1992	PG	1.8 Liter / 4-cylinder	Digifant MFI	102
SLC				
1992–1994	AAA	2.8 Liter / 6-cylinder (VR6)	Motronic	52
EuroVan				
1992–1994	AAF	2.51 Liter / 5-cylinder	Digifant MFI	55
Fox				
1987–1990	UM/JN	1.8 Liter / 4-cylinder	CIS-E	90
1991–1993	ABG	1.8 Liter / 4-cylinder	Digifant	62
Golf				
1985–1987	GX	1.8 Liter / 4-cylinder	CIS or CIS-E	81
1987–1992	RV/PF	1.8 Liter / 4-cylinder	Digifant MFI	112/99
1993–1997	ABA	2.0 Liter / 4-cylinder (8-valve)	Motronic	58
GT				
1986–1988	RD	1.8 Liter / 4-cylinder	CIS-E	109
GTI				
1985–1986	HT	1.8 Liter / 4-cylinder	CIS-E	84
1986–1988	RD	1.8 Liter / 4-cylinder	CIS-E	109
1987–1989	PL	1.8 Liter / 4-cylinder(16-valve)	CIS-E	106
1990–1992	9A	2.0 Liter / 4-cylinder (16-valve)	CIS-E Motronic	65
1991–1992	RV/PF	1.8 Liter / 4-cylinder	Digifant MFI	112/99
1994–1997	AAA	2.8 Liter / 6-cylinder (VR6)	Motronic	52
Jetta (continued on back cover)				
1981–1984	EN	1.7 Liter / 4-cylinder	CIS with Lambda regulation	78
1984	JH	1.8 Liter / 4-cylinder	CIS with Lambda regulation	87
1985–1987	GX	1.8 Liter / 4-cylinder	CIS or CIS-E	81
1987–1992	RV/PF	1.8 Liter / 4-cylinder	Digifant MFI	112/99
1993–1997	ABA	2.0 Liter / 4-cylinder	Motronic	58